Characterization of Antibody Responses to Virus Infections in Humans

Characterization of Antibody Responses to Virus Infections in Humans

Editors

Philipp A. Ilinykh
Kai Huang

MDPI • Basel • Beijing • Wuhan • Barcelona • Belgrade • Manchester • Tokyo • Cluj • Tianjin

Editors
Philipp A. Ilinykh
University of Texas Medical Branch
USA

Kai Huang
University of Texas Medical Branch
USA

Editorial Office
MDPI
St. Alban-Anlage 66
4052 Basel, Switzerland

This is a reprint of articles from the Special Issue published online in the open access journal *Pathogens* (ISSN 2076-0817) (available at: https://www.mdpi.com/journal/pathogens/special_issues/Antibody_Responses_Virus_Infections).

For citation purposes, cite each article independently as indicated on the article page online and as indicated below:

LastName, A.A.; LastName, B.B.; LastName, C.C. Article Title. *Journal Name* **Year**, *Volume Number*, Page Range.

ISBN 978-3-0365-4367-3 (Hbk)
ISBN 978-3-0365-4368-0 (PDF)

Contents

Editorial

What Do Antibody Studies Tell Us about Viral Infections?

Philipp A. Ilinykh [1,2,*] and Kai Huang [1,2,†,‡]

1   Department of Pathology, University of Texas Medical Branch, Galveston, TX 77555, USA; kahuang@utmb.edu
2   Galveston National Laboratory, University of Texas Medical Branch, Galveston, TX 77550, USA
*   Correspondence: pailinyk@utmb.edu
†   Present address: One Health Laboratory, University of Texas Medical Branch, Galveston, TX 77555, USA.
‡   Present address: Department of Internal Medicine (Infectious Diseases), University of Texas Medical Branch, Galveston, TX 77555, USA.

Citation: Ilinykh, P.A.; Huang, K. What Do Antibody Studies Tell Us about Viral Infections? *Pathogens* **2022**, *11*, 560. https://doi.org/10.3390/pathogens11050560

Received: 4 May 2022
Accepted: 6 May 2022
Published: 10 May 2022

Publisher's Note: MDPI stays neutral with regard to jurisdictional claims in published maps and institutional affiliations.

Humoral immunity is an important body defense system against virus infection and is correlated to patient health status. Antibody response is a key factor in controlling virus replication. During infection, viruses induce the production of antibodies, which differ in their isotype, neutralization capacity and breadth, recognition of surface versus internal viral proteins, and epitope specificity. These and other yet to be identified factors determine the role of antibodies in virus clearance through the direct neutralization and Fc effector functions, such as antibody-dependent cellular cytotoxicity [1]. However, certain features of the antibody response, such as antibody-dependent enhancement of infection [2] or increased inflammation resulting from the deposition of immune complexes [3], can create "adverse effects" to exacerbate infection. Adding to the complexity of interaction between viruses and host immune systems, some viruses have exploited multiple mechanisms to compromise antibody production, which helps them to overcome the resistance of host organisms and establish infection. This phenomenon often contributes to the differences in magnitude and longevity of the humoral response to natural infections, in comparison with vaccines [4]. Despite recent advancements in the characterization of monoclonal antibody responses to a number of human pathogens, including human immunodeficiency virus 1 [5], influenza virus [6], dengue virus [7], chikungunya virus [8], rabies virus [9], paramyxoviruses [10], poxviruses [11], hantaviruses [12], filoviruses [13], and coronaviruses [14], critical knowledge gaps still exist. In particular, many viral and host factors that determine the dynamics of antibody response and their role in pathogenesis, as well as the mechanisms of antiviral and proviral antibody effects, remain undefined. Undoubtedly, this information will be vital to guide the design of vaccines and therapeutic strategies based on passive immunization.

The Special Issue "Characterization of Antibody Responses to Virus Infections in Humans" has gathered nine publications, including seven original articles and two reviews, that emphasize the need for better understanding of biological aspects of humoral immune response to different viral pathogens.

The varicella-zoster virus (VZV) belongs to the *Herpesviridae* family and is the causative agent of varicella (chickenpox) and herpes zoster (shingles). After primary replication in the upper respiratory tract, VZV is transported via the bloodstream to the skin sites, causing a widely distributed vesicular rash. VZV can further reach ganglia by axonal transport and establish a latent infection in the nervous system. In case of infection reactivation, the virus is transported down the nerve to the correlating dermatome, which results in zoster. Due to VZV neurotropism, the infection can provoke long-lasting postherpetic neuralgia, especially in elderly and immunocompromised individuals [15,16]. Availability of accurate methods for serodiagnostics of VZV-specific antibodies is needed for the timely treatment of clinical cases, implementation of quarantine measures, vaccination effectiveness control and routine epidemiological surveillance of VZV. As an alternative to commercial kits that are not evenly distributed worldwide, Kombe et al. [17] developed the highly sensitive diagnostic approach based on the chemiluminescent immunoassay, which can detect very low IgA, IgG and IgM titers against VZV-gE envelope glycoprotein in patients at the early stage of infection.

Influenza A viruses (IAV) constitute a large group of pathogens with high relevance for public health. IAVs have been shown to infect humans, pigs, horses, dogs, cats and sea mammals [18–23]. Wild waterfowl serve as a natural reservoir for the vast majority of IAV serotypes. In general, human IAVs cause seasonal flu outbreaks worldwide, with mild-to-severe respiratory symptoms. However, due to the segmented nature of the IAV genome, new viruses emerge as a result of genome reassortment in humans and animals. Given the lack of immunity to such viruses in the human population, these new variants have the potential to cause a pandemic with a high case-fatality ratio [18]. In addition, multiple cases of human infection with avian IAV, predominately the H5 subtype, have been described [24–26] since the first documented outbreak in Hong Kong in 1997 [27–29]. In severe cases, the infection is characterized by excessive lung inflammation resulting from the virus-induced cytokine storm, and can often be fatal [30,31]. Therefore, the serosurveillance studies in 'hot areas', such as South-East Asia, are critical to track the circulation, emergence and evolution of avian IAV to inform outbreak preparedness and response measures. Ilyicheva et al. [32] analyzed serum samples from Vietnamese residents and reported the detection of neutralizing antibodies to H5 avian IAV isolated in Vietnam and Russia in 2017–2018. These findings suggest an ongoing adaptation of the rapidly evolving H5 viruses to human hosts.

The most recent pandemic of viral disease has been caused by severe acute respiratory syndrome coronavirus-2 (SARS-CoV-2), a zoonotic pathogen that barely requires a special introduction nowadays. The World Health Organization declared the COVID-19 outbreak a Public Health Emergency of International Concern on 30 January 2020, and a pandemic on 11 March 2020. It spread rapidly around the world, causing more than 511 million cases and 6.2 million deaths as of 3 May 2022 [33]. The early clinical studies of COVID-19 in China revealed important characteristics of the disease [34,35]. The patients had pneumonia, and, in severe cases, developed acute respiratory distress syndrome and required oxygen therapy in intensive care units. Other common complications included acute cardiac injury and secondary infection. The leucopenia, lymphopenia and high serum levels of proinflammatory cytokines were identified as markers of disease severity. Since the last quarter of 2020, variant viruses have emerged in many parts of the world because of the high burden of infection and the adaptation of SARS-CoV-2 to human cells under immune pressure [36,37].

In this Special Issue, five different publications [38–42] have characterized the antibody response at population and molecular levels, contributing to a broader picture of SARS-CoV-2 epidemiology. Xiao et al. [38] conducted a large-scale screening of serum samples in the Guangdong province, China, between March and June 2020. The overall prevalence of virus-specific antibodies was low soon after the emergence of COVID-19 in Guangdong, suggesting an urgent need for vaccination to increase population immunity to SARS-CoV-2. Another study by Xu et al. [39] revealed a 4.5-fold increase in SARS-CoV-2 seroprevalence from Fall 2020 to February 2021 in the population of Western Pennsylvania, USA, which was shown to be driven both by infection and vaccination. Kazachinskaia et al. [40] analyzed the blood samples of COVID-19 patients in Novosibirsk city, Russia, and observed cross-reactivity of antibodies to SARS-CoV (2002) proteins. Additionally, their results suggested that high neutralization titer is not necessarily predictive of the infection survival. Huang et al. [41] presented the kinetics study of viral load, humoral immune response and the cytokine profile in a hospital patient cohort (January–March 2020) and were able to correlate these parameters with the disease severity during the initial outbreak in Taiwan. Another work from Germany by Heidepriem et al. [42] provided longitudinal characterization of the antibody response to SARS-CoV-2. The linear epitopes in viral proteome and specific glycan structures, targeted by antibodies from COVID-19 patients with moderate and mild disease, were identified.

Filoviruses represent yet another group of pathogens with high relevance for global health and include one of the deadliest human pathogens known so far, Ebola (EBOV) and Marburg viruses. These and other members of the *Filoviridae* family can cause a severe disease, which is often accompanied by haemorrhagic manifestations and systemic multiorgan dysfunction, with fatality rates reaching as high as 90% [43]. Human outbreaks generally result from spillover events from infected animals, including bats and non-human primates [44]. The development of infection is believed to result from deep suppression of the host immune system and dysregulation of both innate and adaptive arms of immunity by filoviruses. In worst cases, the rapid disease progression culminates in the death of an infected individual in 1–2 weeks after the onset of symptoms. The largest epidemic of filovirus-induced disease occurred in 2013–2016 in West Africa and claimed the lives of 11,310 out of 28,616 people infected [45].

The filovirus glycoprotein (GP) is the sole envelope viral protein responsible for cell entry; hence, it serves as the primary target for antibody-based therapies and vaccine design efforts. Currently, monoclonal antibody (mAb) therapy has been shown to be the most effective treatment of filoviral infections after the onset of symptoms [46]. In this Special Issue, two comprehensive review papers by Hargreaves et al. [47] and Yu and Saphire [48] summarize the recent advancements in the characterization of neutralizing antibody responses against EBOV and other filoviruses. The authors discuss the role of epitope specificity and Fc effector functions in antiviral mechanisms employed by the most promising antibodies, the correlation of these parameters with in vivo protection by individual mAbs and mAb cocktails, the structural basis for cross-reactivity to ebolavirus species and the strategies to avoid viral escape from neutralizing antibodies.

In conclusion, we believe that this Special Issue underlines the importance of multi-level analysis of antibody responses in the context of virus infections. The data presented here contribute to a better understanding of epidemiological and molecular aspects of infectious diseases caused by publicly relevant viral pathogens, such as VZV, IAV, SARS-CoV-2 and filoviruses. We hope that this Special Issue will stimulate future studies on humoral immune response to inform the development of countermeasures against life-threatening viral infections.

Acknowledgments: We thank Michelle N. Meyer (UTMB) for the critical reading of the manuscript and useful suggestions.

References

1. Murphy, K.; Weaver, C. *Janeway's Immunobiology*; Garland Science: New York, NY, USA, 2017.
2. Taylor, A.; Foo, S.S.; Bruzzone, R.; Dinh, L.V.; King, N.J.; Mahalingam, S. Fc receptors in antibody-dependent enhancement of viral infections. *Immunol. Rev.* **2015**, *268*, 340–364. [CrossRef] [PubMed]
3. Diamond, M.S.; Pierson, T.C. The Challenges of Vaccine Development against a New Virus during a Pandemic. *Cell Host Microbe* **2020**, *27*, 699–703. [CrossRef] [PubMed]
4. Ilinykh, P.A.; Bukreyev, A. Antibody responses to filovirus infections in humans: Protective or not? *Lancet Infect. Dis.* **2021**, *21*, e348–e355. [CrossRef]
5. Walker, L.M.; Phogat, S.K.; Chan-Hui, P.Y.; Wagner, D.; Phung, P.; Goss, J.L.; Wrin, T.; Simek, M.D.; Fling, S.; Mitcham, J.L.; et al. Broad and potent neutralizing antibodies from an African donor reveal a new HIV-1 vaccine target. *Science* **2009**, *326*, 285–289. [CrossRef] [PubMed]
6. Corti, D.; Voss, J.; Gamblin, S.J.; Codoni, G.; Macagno, A.; Jarrossay, D.; Vachieri, S.G.; Pinna, D.; Minola, A.; Vanzetta, F.; et al. A neutralizing antibody selected from plasma cells that binds to group 1 and group 2 influenza A hemagglutinins. *Science* **2011**, *333*, 850–856. [CrossRef] [PubMed]
7. Dejnirattisai, W.; Wongwiwat, W.; Supasa, S.; Zhang, X.; Dai, X.; Rouvinski, A.; Jumnainsong, A.; Edwards, C.; Quyen, N.T.H.; Duangchinda, T.; et al. A new class of highly potent, broadly neutralizing antibodies isolated from viremic patients infected with dengue virus. *Nat. Immunol.* **2015**, *16*, 170–177. [CrossRef]

8. Smith, S.A.; Silva, L.A.; Fox, J.M.; Flyak, A.I.; Kose, N.; Sapparapu, G.; Khomandiak, S.; Ashbrook, A.W.; Kahle, K.M.; Fong, R.H.; et al. Isolation and Characterization of Broad and Ultrapotent Human Monoclonal Antibodies with Therapeutic Activity against Chikungunya Virus. *Cell Host Microbe* **2015**, *18*, 86–95. [CrossRef]

9. De Benedictis, P.; Minola, A.; Rota Nodari, E.; Aiello, R.; Zecchin, B.; Salomoni, A.; Foglierini, M.; Agatic, G.; Vanzetta, F.; Lavenir, R.; et al. Development of broad-spectrum human monoclonal antibodies for rabies post-exposure prophylaxis. *EMBO Mol. Med.* **2016**, *8*, 407–421. [CrossRef]

10. Corti, D.; Bianchi, S.; Vanzetta, F.; Minola, A.; Perez, L.; Agatic, G.; Guarino, B.; Silacci, C.; Marcandalli, J.; Marsland, B.J.; et al. Cross-neutralization of four paramyxoviruses by a human monoclonal antibody. *Nature* **2013**, *501*, 439–443. [CrossRef]

11. Gilchuk, I.; Gilchuk, P.; Sapparapu, G.; Lampley, R.; Singh, V.; Kose, N.; Blum, D.L.; Hughes, L.J.; Satheshkumar, P.S.; Townsend, M.B.; et al. Cross-Neutralizing and Protective Human Antibody Specificities to Poxvirus Infections. *Cell* **2016**, *167*, 684–694.e689. [CrossRef]

12. Engdahl, T.B.; Kuzmina, N.A.; Ronk, A.J.; Mire, C.E.; Hyde, M.A.; Kose, N.; Josleyn, M.D.; Sutton, R.E.; Mehta, A.; Wolters, R.M.; et al. Broad and potently neutralizing monoclonal antibodies isolated from human survivors of New World hantavirus infection. *Cell Rep.* **2021**, *36*, 109453. [CrossRef] [PubMed]

13. Saphire, E.O.; Schendel, S.L.; Fusco, M.L.; Gangavarapu, K.; Gunn, B.M.; Wec, A.Z.; Halfmann, P.J.; Brannan, J.M.; Herbert, A.S.; Qiu, X.; et al. Systematic Analysis of Monoclonal Antibodies against Ebola Virus GP Defines Features that Contribute to Protection. *Cell* **2018**, *174*, 938–952.e913. [CrossRef] [PubMed]

14. Hastie, K.M.; Li, H.; Bedinger, D.; Schendel, S.L.; Dennison, S.M.; Li, K.; Rayaprolu, V.; Yu, X.; Mann, C.; Zandonatti, M.; et al. Defining variant-resistant epitopes targeted by SARS-CoV-2 antibodies: A global consortium study. *Science* **2021**, *374*, 472–478. [CrossRef] [PubMed]

15. Zerboni, L.; Sen, N.; Oliver, S.L.; Arvin, A.M. Molecular mechanisms of varicella zoster virus pathogenesis. *Nat. Rev. Microbiol.* **2014**, *12*, 197–210. [CrossRef]

16. Patil, A.; Goldust, M.; Wollina, U. Herpes zoster: A Review of Clinical Manifestations and Management. *Viruses* **2022**, *14*, 192. [CrossRef]

17. Kombe Kombe, A.J.; Xie, J.; Zahid, A.; Ma, H.; Xu, G.; Deng, Y.; Nsole Biteghe, F.A.; Mohammed, A.; Dan, Z.; Yang, Y.; et al. Detection of Circulating VZV-Glycoprotein E-Specific Antibodies by Chemiluminescent Immunoassay (CLIA) for Varicella-Zoster Diagnosis. *Pathogens* **2022**, *11*, 66. [CrossRef]

18. Webster, R.G.; Bean, W.J.; Gorman, O.T.; Chambers, T.M.; Kawaoka, Y. Evolution and ecology of influenza A viruses. *Microbiol Rev.* **1992**, *56*, 152–179. [CrossRef]

19. Ma, W. Swine influenza virus: Current status and challenge. *Virus Res.* **2020**, *288*, 198118. [CrossRef]

20. Chambers, T.M. Equine Influenza. *Cold Spring Harb. Perspect. Med.* **2022**, *12*, a038331. [CrossRef]

21. Lyu, Y.; Song, S.; Zhou, L.; Bing, G.; Wang, Q.; Sun, H.; Chen, M.; Hu, J.; Wang, M.; Sun, H.; et al. Canine Influenza Virus A(H3N2) Clade with Antigenic Variation, China, 2016–2017. *Emerg. Infect. Dis.* **2019**, *25*, 161–165. [CrossRef]

22. Frymus, T.; Belak, S.; Egberink, H.; Hofmann-Lehmann, R.; Marsilio, F.; Addie, D.D.; Boucraut-Baralon, C.; Hartmann, K.; Lloret, A.; Lutz, H.; et al. Influenza Virus Infections in Cats. *Viruses* **2021**, *13*, 1435. [CrossRef] [PubMed]

23. Runstadler, J.A.; Puryear, W. A Brief Introduction to Influenza A Virus in Marine Mammals. *Methods Mol. Biol.* **2020**, *2123*, 429–450. [CrossRef] [PubMed]

24. Sutton, T.C. The Pandemic Threat of Emerging H5 and H7 Avian Influenza Viruses. *Viruses* **2018**, *10*, 461. [CrossRef] [PubMed]

25. de Jong, M.D.; Hien, T.T. Avian influenza A (H5N1). *J. Clin. Virol.* **2006**, *35*, 2–13. [CrossRef] [PubMed]

26. Neumann, G.; Chen, H.; Gao, G.F.; Shu, Y.; Kawaoka, Y. H5N1 influenza viruses: Outbreaks and biological properties. *Cell Res.* **2010**, *20*, 51–61. [CrossRef]

27. Li, K.S.; Guan, Y.; Wang, J.; Smith, G.J.; Xu, K.M.; Duan, L.; Rahardjo, A.P.; Puthavathana, P.; Buranathai, C.; Nguyen, T.D.; et al. Genesis of a highly pathogenic and potentially pandemic H5N1 influenza virus in eastern Asia. *Nature* **2004**, *430*, 209–213. [CrossRef]

28. Chen, H.; Smith, G.J.; Li, K.S.; Wang, J.; Fan, X.H.; Rayner, J.M.; Vijaykrishna, D.; Zhang, J.X.; Zhang, L.J.; Guo, C.T.; et al. Establishment of multiple sublineages of H5N1 influenza virus in Asia: Implications for pandemic control. *Proc. Natl. Acad. Sci. USA* **2006**, *103*, 2845–2850. [CrossRef]

29. Smith, G.J.; Fan, X.H.; Wang, J.; Li, K.S.; Qin, K.; Zhang, J.X.; Vijaykrishna, D.; Cheung, C.L.; Huang, K.; Rayner, J.M.; et al. Emergence and predominance of an H5N1 influenza variant in China. *Proc. Natl. Acad. Sci. USA* **2006**, *103*, 16936–16941. [CrossRef]

30. de Jong, M.D.; Simmons, C.P.; Thanh, T.T.; Hien, V.M.; Smith, G.J.; Chau, T.N.; Hoang, D.M.; Chau, N.V.; Khanh, T.H.; Dong, V.C.; et al. Fatal outcome of human influenza A (H5N1) is associated with high viral load and hypercytokinemia. *Nat. Med.* **2006**, *12*, 1203–1207. [CrossRef]

31. Cheung, C.Y.; Poon, L.L.; Lau, A.S.; Luk, W.; Lau, Y.L.; Shortridge, K.F.; Gordon, S.; Guan, Y.; Peiris, J.S. Induction of proinflammatory cytokines in human macrophages by influenza A (H5N1) viruses: A mechanism for the unusual severity of human disease? *Lancet* **2002**, *360*, 1831–1837. [CrossRef]

32. Ilyicheva, T.; Marchenko, V.; Pyankova, O.; Moiseeva, A.; Nhai, T.T.; Lan Anh, B.T.; Sau, T.K.; Kuznetsov, A.; Ryzhikov, A.; Maksyutov, R. Antibodies to Highly Pathogenic A/H5Nx (Clade 2.3.4.4) Influenza Viruses in the Sera of Vietnamese Residents. *Pathogens* **2021**, *10*, 394. [CrossRef] [PubMed]

33. World Health Organization. WHO Coronavirus (COVID-19) Dashboard. Available online: https://covid19.who.int/ (accessed on 7 April 2022).
34. Huang, C.; Wang, Y.; Li, X.; Ren, L.; Zhao, J.; Hu, Y.; Zhang, L.; Fan, G.; Xu, J.; Gu, X.; et al. Clinical features of patients infected with 2019 novel coronavirus in Wuhan, China. *Lancet* **2020**, *395*, 497–506. [CrossRef]
35. Zhang, X.; Tan, Y.; Ling, Y.; Lu, G.; Liu, F.; Yi, Z.; Jia, X.; Wu, M.; Shi, B.; Xu, S.; et al. Viral and host factors related to the clinical outcome of COVID-19. *Nature* **2020**, *583*, 437–440. [CrossRef] [PubMed]
36. Subbarao, K. The success of SARS-CoV-2 vaccines and challenges ahead. *Cell Host Microbe* **2021**, *29*, 1111–1123. [CrossRef] [PubMed]
37. Andreano, E.; Rappuoli, R. SARS-CoV-2 escaped natural immunity, raising questions about vaccines and therapies. *Nat. Med.* **2021**, *27*, 759–761. [CrossRef]
38. Xiao, C.; Leung, N.H.L.; Cheng, Y.; Lei, H.; Ling, S.; Lin, X.; Tao, R.; Huang, X.; Guan, W.; Yang, Z.; et al. Seroprevalence of Antibodies to SARS-CoV-2 in Guangdong Province, China between March to June 2020. *Pathogens* **2021**, *10*, 1505. [CrossRef]
39. Xu, L.; Doyle, J.; Barbeau, D.J.; Le Sage, V.; Wells, A.; Duprex, W.P.; Shurin, M.R.; Wheeler, S.E.; McElroy, A.K. A Cross-Sectional Study of SARS-CoV-2 Seroprevalence between Fall 2020 and February 2021 in Allegheny County, Western Pennsylvania, USA. *Pathogens* **2021**, *10*, 710. [CrossRef]
40. Kazachinskaia, E.; Chepurnov, A.; Shcherbakov, D.; Kononova, Y.; Saroyan, T.; Gulyaeva, M.; Shanshin, D.; Romanova, V.; Khripko, O.; Voevoda, M.; et al. IgG Study of Blood Sera of Patients with COVID-19. *Pathogens* **2021**, *10*, 1421. [CrossRef]
41. Huang, C.G.; Dutta, A.; Huang, C.T.; Chang, P.Y.; Hsiao, M.J.; Hsieh, Y.C.; Lin, S.M.; Shih, S.R.; Tsao, K.C.; Yang, C.T. Relative COVID-19 Viral Persistence and Antibody Kinetics. *Pathogens* **2021**, *10*, 752. [CrossRef]
42. Heidepriem, J.; Dahlke, C.; Kobbe, R.; Santer, R.; Koch, T.; Fathi, A.; Seco, B.M.S.; Ly, M.L.; Schmiedel, S.; Schwinge, D.; et al. Longitudinal Development of Antibody Responses in COVID-19 Patients of Different Severity with ELISA, Peptide, and Glycan Arrays: An Immunological Case Series. *Pathogens* **2021**, *10*, 438. [CrossRef]
43. Feldmann, H.; Sanchez, A.; Geisbert, T. Filoviridae: Marburg and Ebola Viruses. In *Fields Virology*, 6th ed.; Wolters Kluwer Health Adis (ESP): London, UK, 2013; Volume 1.
44. Feldmann, H.; Sprecher, A.; Geisbert, T.W. Ebola. *N. Engl. J. Med.* **2020**, *382*, 1832–1842. [CrossRef] [PubMed]
45. Centers for Disease Control and Prevention. Number of Cases and Deaths in Guinea, Liberia, and Sierra Leone during the 2014–2016 West Africa Ebola Outbreak. Available online: https://www.cdc.gov/vhf/ebola/outbreaks/2014-west-africa/case-counts.html (accessed on 7 April 2022).
46. Mulangu, S.; Dodd, L.E.; Davey, R.T., Jr.; Tshiani Mbaya, O.; Proschan, M.; Mukadi, D.; Lusakibanza Manzo, M.; Nzolo, D.; Tshomba Oloma, A.; Ibanda, A.; et al. A Randomized, Controlled Trial of Ebola Virus Disease Therapeutics. *N. Engl. J. Med.* **2019**, *381*, 2293–2303. [CrossRef] [PubMed]
47. Hargreaves, A.; Brady, C.; Mellors, J.; Tipton, T.; Carroll, M.W.; Longet, S. Filovirus Neutralising Antibodies: Mechanisms of Action and Therapeutic Application. *Pathogens* **2021**, *10*, 1201. [CrossRef] [PubMed]
48. Yu, X.; Saphire, E.O. Development and Structural Analysis of Antibody Therapeutics for Filoviruses. *Pathogens* **2022**, *11*, 374. [CrossRef]

Article

Detection of Circulating VZV-Glycoprotein E-Specific Antibodies by Chemiluminescent Immunoassay (CLIA) for Varicella–Zoster Diagnosis

Arnaud John Kombe Kombe [1], Jiajia Xie [1], Ayesha Zahid [2], Huan Ma [2], Guangtao Xu [2], Yiyu Deng [2], Fleury Augustin Nsole Biteghe [3,4], Ahmed Mohammed [2], Zhao Dan [2], Yunru Yang [2], Chen Feng [2], Weihong Zeng [2], Ruixue Chang [1], Keyuan Zhu [1], Siping Zhang [1,*] and Tengchuan Jin [1,2,5,*]

[1] Department of Dermatology, The First Affiliated Hospital of USTC, Division of Life Sciences and Medicine, University of Science and Technology of China, Hefei 230001, China; kombe@mail.ustc.edu.cn (A.J.K.K.); xiejiajia@ustc.edu.cn (J.X.); crx0419@163.com (R.C.); zhukeyuan1106@163.com (K.Z.)

[2] Hefei National Laboratory for Physical Sciences at Microscale, The CAS Key Laboratory of Innate Immunity and Chronic Disease, School of Basic Medical Sciences, Division of Life Sciences and Medicine, University of Science and Technology of China, Hefei 230026, China; ayesha1@mail.ustc.edu.cn (A.Z.); mahuan86@ustc.edu.cn (H.M.); taotao222@mail.ustc.edu.cn (G.X.); pb18081629@mail.ustc.edu.cn (Y.D.); amam@mail.ustc.edu.cn (A.M.); zhdan123@mail.ustc.edu.cn (Z.D.); yyr@mail.ustc.edu.cn (Y.Y.); cf155@mail.ustc.edu.cn (C.F.); whongz@ustc.edu.cn (W.Z.)

[3] Gabonese Scientific Research Consortium, Libreville, Gabon; Fleury.nsolebiteghe@cshs.org

[4] Department of Radiation Oncology, Cedars Sinai Hospital, Los Angeles, CA 90048, USA

[5] CAS Center for Excellence in Molecular Cell Science, Chinese Academy of Sciences, Shanghai 200031, China

[*] Correspondence: zh_siping@126.com (S.Z.); jint@ustc.edu.cn (T.J.); Tel.: +86-0551-62283151 (S.Z.); +86-551-63600720 (T.J.)

Citation: Kombe Kombe, A.J.; Xie, J.; Zahid, A.; Ma, H.; Xu, G.; Deng, Y.; Nsole Biteghe, F.A.; Mohammed, A.; Dan, Z.; Yang, Y.; et al. Detection of Circulating VZV-Glycoprotein E-Specific Antibodies by Chemiluminescent Immunoassay (CLIA) for Varicella–Zoster Diagnosis. *Pathogens* **2022**, *11*, 66. https://doi.org/10.3390/pathogens11010066

Academic Editors: Philipp A. Ilinykh, Kai Huang and Lawrence S. Young

Received: 11 October 2021
Accepted: 23 November 2021
Published: 5 January 2022

Publisher's Note: MDPI stays neutral with regard to jurisdictional claims in published maps and institutional affiliations.

Abstract: Varicella and herpes zoster are mild symptoms-associated diseases caused by varicella–zoster virus (VZV). They often cause severe complications (disseminated zoster), leading to death when diagnoses and treatment are delayed. However, most commercial VZV diagnostic tests have low sensitivity, and the most sensitive tests are unevenly available worldwide. Here, we developed and validated a highly sensitive VZV diagnostic kit based on the chemiluminescent immunoassay (CLIA) approach. VZV-glycoprotein E (gE) was used to develop a CLIA diagnostic approach for detecting VZV-specific IgA, IgG, and IgM. The kit was tested with 62 blood samples from 29 VZV-patients classified by standard ELISA into true-positive and equivocal groups and 453 blood samples from VZV-negative individuals. The diagnostic accuracy of the CLIA kit was evaluated by receiver-operating characteristic (ROC) analysis. The relationships of immunoglobulin-isotype levels between the two groups and with patient age ranges were analyzed. Overall, the developed CLIA-based diagnostic kit demonstrated the detection of VZV-specific immunoglobulin titers depending on sample dilution. From the ELISA-based true-positive patient samples, the diagnostic approach showed sensitivities of 95.2%, 95.2%, and 97.6% and specificities of 98.0%, 100%, and 98.9% for the detection of VZV-gE-specific IgA, IgG, and IgM, respectively. Combining IgM to IgG and IgA detection improved diagnostic accuracy. Comparative analyses on diagnosing patients with equivocal results displaying very low immunoglobulin titers revealed that the CLIA-based diagnostic approach is overall more sensitive than ELISA. In the presence of typical VZV symptoms, CLIA-based detection of high titer of IgM and low titer of IgA/IgG suggested the equivocal patients experienced primary VZV infection. Furthermore, while no difference in IgA/IgG level was found regarding patient age, IgM level was significantly higher in young adults. The CLIA approach-based detection kit for diagnosing VZV-gE-specific IgA, IgG, and IgM is simple, suitable for high-throughput routine analysis situations, and provides enhanced specificity compared to ELISA.

Keywords: varicella–zoster virus (VZV); chemiluminescent immunoassay (CLIA); IgA; IgG; IgM; diagnostic test

1. Introduction

Varicella–zoster virus (VZV), also known as human herpesvirus-3 (HHV-3), belongs to the α-herpesviridae subfamily [1]. It is responsible for varicella disease or chickenpox in children, adolescents, and young adults. The latent viral resurgence years later commonly occurs in older people and causes a secondary infection known as zoster or shingles [1,2].

Although considered among the mild-symptom diseases, VZV-related diseases are highly morbid. The most life-threatening complications include mental development deficit, meningoencephalitis, and post-infectious encephalopathy, in varicella cases. In herpes zoster cases, complications include vasculitis, zoster sine herpete, and post-herpetic neuralgia. A lack of early diagnosis results in treatment delays, which usually leads to fatalities, especially in newborns, elders, organ transplant recipients, and immunocompromised people experiencing disseminated herpes zoster [3–8].

Nowadays, VZV vaccination has led to a significant decrease in the incidence of varicella, particularly in countries where vaccination programs have been implemented and well followed [9–11]. Consequently, this has decreased hospitalization and remarkably reduced the routine biological diagnoses in laboratories [12,13]. In these countries, VZV immunodiagnostic tests assessing IgA, IgG, and/or IgM have been only recommended in pregnant women, critically ill patients before organ transplant surgery, and immunocompromised people, post-vaccinated people, and hospital practitioners [14,15]. However, many reports hypothesized that the implementation of varicella vaccines would be followed by an increase in post-vaccine varicella or herpes zoster cases [2,9,10,16–18], suggesting the need for using antibody detection tests in routine diagnoses for global epidemiologic surveillance, along with herpes zoster vaccination [17]. For instance, in many other countries, such as China [18–20] and Norway [21], where anti-VZV vaccines are not yet implemented or where VZV vaccination coverage is uneven, rapid case identifications are crucial [19,21,22]. Reporting varicella cases in these countries, especially in high-frequented public areas such as schools, institutions, healthcare centers, hospitals, etc., would prevent rapid infection spread to people at risk (pregnant women, immunocompromised). Consequently, this will prevent them from progressing toward infection complication stages and facilitate outbreak control, as diagnosis delays are often fatal.

While the diagnosis of VZV infection is needed in both countries with well-established and non-implemented VZV-vaccine programs, routine biological diagnostics has become challenging, as many currently available diagnostic tests have low sensitivity/specificity [9,13,23]. The few highly sensitive immune diagnostic tests are scarce in the market or not evenly available worldwide [13]. Practically, in the past decade, several biological diagnostic tests with variable sensitivities have been developed to detect VZV-specific IgA, IgG, or IgM, or polyclonal antibodies. Most of them, including direct fluorescent antibody (DFA) and Tzanck smear diagnostic kits based on immunofluorescent assay (IFA), yield in low to moderate antibody detection sensitivity, around 60–80%, and 42–90%, respectively [24–26]. The VZV detection using virus culture assays resulted in high toxicities and contaminations, biasing the diagnostic results, as yielded mainly in false-negative (46% of sensitivity) [25]. In addition, these immunodiagnostic approaches are generally labor-intensive and time consuming, requiring meticulous specimen collection and highly trained technicians [25]. Those with high detection sensitivities and specificities (around 97.8% and 96.8%, respectively), including glycoprotein-based enzyme-linked immunosorbent assay (gpELISA) or varicella zoster glycoprotein IgG enzyme immunoassay with a reference time-resolved fluorescence immunoassay (VZV TRFIA), VZV IgG glycoprotein assay (Merck gpEIA) for the detection of serum VZV IgG, are not evenly and/or commercially available worldwide [27].

Although molecular diagnostics, including viral isolation from vesicular fluid cultures and swab samples, and nucleic acid detection (by PCR), are the most sensitive in VZV diagnostic [24,28], antibody assessment is needed in VZV epidemiological surveillance [11,12] and vaccination effectiveness control [9]. For instance, vesicular rashes do not always appear during infection, compromising PCR results [29,30]. Moreover, the main diagnostic approach based on clinical presentation is not 100% reliable, since many other herpetic diseases, including HSV, present similarly. Therefore, there is still a need for widely commercially distributed tests with high sensitivity and specificity for routine VZV diagnostics.

Several studies support that the chemiluminescent immunoassay (CLIA) approach for diagnosing serum/plasma viral antigen-specific IgA, IgG, or IgM has better diagnostic performance among the known immunoassays, including manual ELISA [31–33]. Manual ELISA is a labor-intensive multiple-wash-based assay; therefore, it is not suitable for high-throughput screening situations. Moreover, ELISA hardly detects weak antibody–antigen interactions and results in high background, compromising the test sensitivity. Interestingly, in the context of routine diagnostic of SARS-CoV-2, we and others [34,35] have developed CLIA-based diagnostic methods currently commercially available and with satisfying added values compared to ELISA [35] for diagnosing and monitoring COVID-19.

Therefore, we developed and validated a highly sensitive and specific diagnostic kit based on the CLIA approach for diagnosing VZV infections.

The developed CLIA-based VZV diagnostic approach demonstrated improved diagnostic accuracy, as it could detect very low IgA, IgG, and IgM titer in patients at the early stage of the VZV infection. Moreover, as the diagnosis process is automated, time-saving, and suitable for high-throughput situations, it can be used for routine diagnoses of VZV infections.

2. Results

2.1. Patient Characteristics and Sampling

Overall, 29 people (Supplementary Figure S1) with an average age of 52 (20 to 82 years old) were enrolled and retained as VZV-infected patients, which was based on typical VZV symptoms and using ELISA tests. None of them was (or has been diagnosed) positive for herpes simplex viruses (HSV1/2), underwent an organ transplant surgery, or received an anti-VZV vaccine. Hepatitis A, B, and E were the only pathologies found among the patients. Supplementary Table S1 describes the clinical and epidemiological characteristics of the included patients.

A total of 62 blood samples from the 29 retained VZV-patients and 453 plasmas/sera from random healthy people (used as negative controls) were obtained to test the developed diagnostic approach.

2.2. Patient Samples React with VZV Glycoprotein Depending on Concentration: Cohort Stratification

As aforementioned, three ELISA kits (ab108781, ab108782, and ab108783 for IgA/IgG/IgM; Abcam) were used to assess the VZV-specific IgA, IgG, and IgM, respectively, in VZV-patient samples from the 29 included VZV-patients. Sixteen samples from random healthy people were used as negative controls. According to the manufacturer's instructions, these ELISA assays were considered standard for VZV-specific antibody detection in patients to confirm and stratify the retained patients as true-positive or equivocal groups. All samples were first serially diluted and assessed for VZV immunoglobulin detection. As a result, all patient samples (but not those from negative controls) reacted with VZV glycoproteins in a concentration-dependent manner (Supplementary Figure S2). These results confirmed the presence of VZV-specific IgA, IgG, and IgM in VZV-patient samples.

Specifically, of the 29 included patients, 21 showed moderate to high reactivity, even at high dilution for IgA, IgG, and IgM detection, respectively. With OD450 $\geq$ 1 at 1/100 dilution, all these 21 patients were considered true-positive for IgA, IgG, and IgM according to the ELISAs' manufacturer instructions (Figure 1, Supplementary Figures S1 and S2).

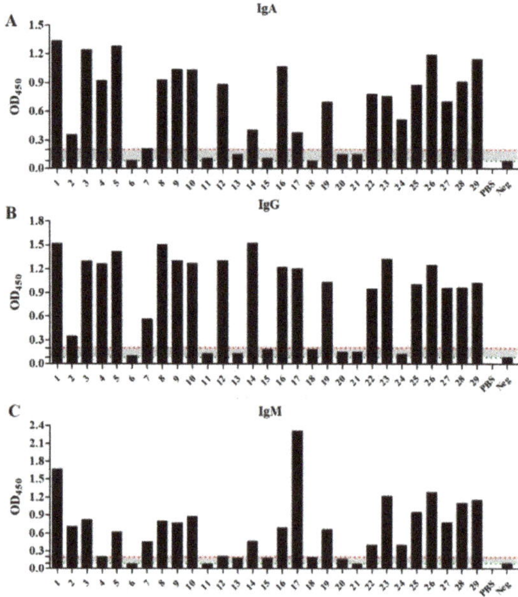

Figure 1. Results for confirmation and stratification of VZV infection in the recruited cohort. ELISA was performed using ab108781 (**A**), ab108782 (**B**), and ab108783 (**C**) ELISA tests, at 1/100 dilution for each sample. Patients with OD450 above 0.2 (red dotted line) were considered positive; with OD450 between 0.1 and 0.2 (the gray area between positive and negative threshold), they were equivocal; and with OD450 below 0.1 (green dotted line), they were considered negatives.

In contrast, eight patient samples (patient number 6, 11, 13, 15, 18, 20, 21, and 24) showed inconsistent results in IgA, IgG, and IgM detection, respectively. Precisely, each patient sample showed equivocal results (OD450 between 0.1 and 0.21 at 1/100 dilution) for at least one of the three antibody isotypes IgA, IgG, and IgM (Figure 1, Supplementary Figure S2). Therefore, they were considered as equivocal for further analyses.

2.3. Anti-VZV-gE IgA, IgG, and IgM Detection-Based VZV Diagnostic Kit Has High Sensitivity/Accuracy

The highly purified VZV-gE antigen (Supplementary Figure S3) was used to make the CLIA-based IgG, IgA, and IgM detection kit, respectively. Then, this developed approach was tested for detecting VZV-gE-specific IgA, IgG, and IgM in sera and plasmas of patients. The reliability of the developed CLIA diagnostic kit was assessed with a two-fold serial dilution of the 62 samples from the 29 patients along with the 16 negative samples used in standard ELISA. As expecting and similar to the standard ELISA results, the CLIA approach showed sample dilution-dependent results (Supplementary Figure S4). These results validated the CLIA diagnostic approach.

Then, to assess the performance of the CLIA approach, the cohort of 42 samples from the 21 true-positive patients and 453 independent samples (plasma or serum) from healthy people were tested by the developed VZV-gE-IgA/IgG/IgM kit. Before testing, samples were pre-treated (virus-inactivated) and diluted 20 times with dilution buffer (PBS) supplemented with 2% BSA. ROC analysis results showed sensitivities of 95.2% (IC95%: 76.2–99.9%), 95.2% (IC95%: 83.8–99.4%), and 97.6% (IC95%: 87.4–99.9%) and specificities of 98.0% (IC95%: 96.3–99.1%), 100% (IC95%: 99.2–100%), and 98.90% (IC95%: 97.4–99.6%) for IgA, IgG, and IgM detection, respectively (Figure 2A–C, Table 1). The cut-offs (criterion) for the IgA, IgG, and IgM diagnostic tests were >78662 RLU, >23450 RLU, and >89634 RLU, respectively (Figure 2A–C).

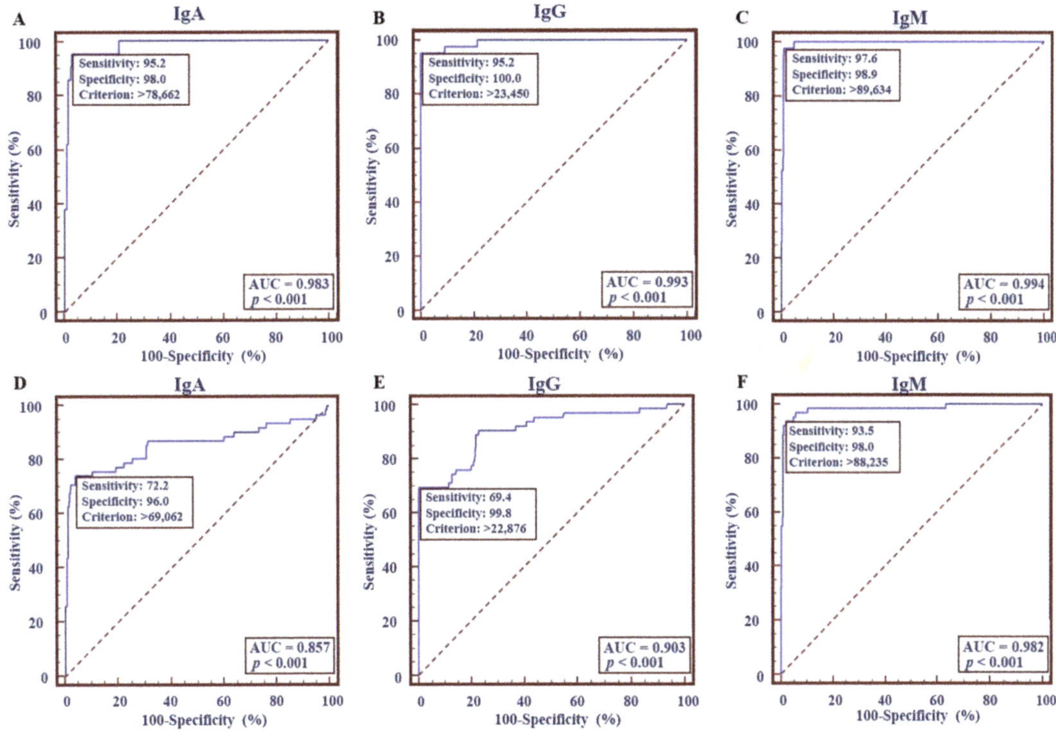

Figure 2. Performance of VZV-gE specific IgA, IgG, and IgM detection kits. The receiver-operating characteristic (ROC) curve analysis for detection of anti-IgA, IgG, and IgM antibodies against VZV-gE protein obtained from 42 ELISA positive samples, regardless (**A–C**), and considering as positive (**D–F**) the equivocal patient samples. The area under the curve (AUC) and the p-value are shown.

Table 1. Sensitivity, specificity, and overall agreements of each VZV-gE-specific IgA, IgG, and IgM kit and their combinations in diagnosing varicella–zoster.

Antibody Type	Sensitivity			Specificity			Overall Agreement	
	n/Total	%	IC95%	n/Total	%	IC95%	n/Total	%
IgA [#]	40/42	95.2	76.2–99.9	444/453	98.0	96.3–99.1	484/495	97.8
IgG [#]	40/42	95.2	83.8–99.4	453/453	100	99.2–100	493/495	99.6
IgM [#]	41/42	97.6	87.4–99.9	448/453	98.9	97.4–99.6	489/495	98.8
IgA [#] & IgG [#]	38/42	90.5	NA	440/453	97.2	NA	480/495	97.0
IgG [#] & IgM [#]	40/42	95.2	NA	448/453	98.9	NA	488/495	98.6
IgA [#] & IgM [#]	40/42	95.2	NA	440/453	97.2	NA	480/495	97.0
IgA [#] & IgG [#] & IgM [#]	39/42	92.9	NA	440/453	97.2	NA	480/495	97.0
IgA [#] or IgG [#]	40/42	95.2	NA	453/453	100	NA	493/495	99.6
IgG [#] or IgM [#]	41/42	97.6	NA	453/453	100	NA	494/495	99.8
IgA [#] or IgM [#]	42/42	100	NA	452/453	99.8	NA	494/495	99.8
IgA [#] or IgG [#] or IgM [#]	41/42	97.6	NA	453/453	100	NA	494/495	99.8
IgA [*]	46/62	74.2	61.5–84.5	435/453	96.0	93.8–97.6	481/515	93.4
IgG [*]	43/62	69.4	56.3–80.4	452/453	99.8	98.0–100	495/515	96.1
IgM [*]	58/62	93.6	84.3–98.2	444/453	98.0	96.3–99.1	502/515	97.5
IgA [*] & IgG [*]	40/62	64.5	NA	434/453	95.8	NA	474/515	92.0
IgG [*] & IgM [*]	42/62	67.7	NA	443/453	97.8	NA	485/515	95.2
IgA [*] & IgM [*]	43/62	69.4	NA	427/453	94.3	NA	470/515	91.3
IgA [*] & IgG [*] & IgM [*]	39/62	62.9	NA	426/453	94.0	NA	465/515	90.3

Table 1. *Cont.*

Antibody Type	Sensitivity			Specificity			Overall Agreement	
	n/Total	%	IC95%	n/Total	%	IC95%	n/Total	%
IgA * or IgG *	49/62	79.0	NA	453/453	100	NA	502/515	97.5
IgG * or IgM *	59/62	95.2	NA	453/453	100	NA	512/515	99.4
IgA * or IgM *	61/62	98.4	NA	452/453	99.8	NA	513/515	99.6
IgA * or IgG * or IgM *	61/62	98.4	NA	453/453	100	NA	514/515	99.8

CLIA-based kit diagnoses features obtained regardless [#] and regarding as positive * the equivocal samples; NA: non-applicable.

2.4. Combining IgM to IgG and IgA Detection Improves the Accuracy of the Varicella–Zoster Diagnosis

It is well known that IgMs are the first immunoglobulins produced at the first contact with viruses. Days after infection, IgM titer in the blood decreases while IgA and IgG titer increases [36]. As previously shown, adding IgA to serological CLIA improves the accuracy of SARS-CoV-2 diagnosis [34]. Then, we assessed whether the detection of IgM combined with that of IgG and/or IgA would be beneficial to the diagnosis of varicella and herpes zoster. As shown in Table 1, when VZV-gE specific IgM detection and one (but interestingly both) of the VZV-gE specific IgG and IgA detection were combined, the sensitivity, specificity, and the overall agreement increased significantly to 97.6%, 100%, and 99.8%, respectively. This combination has higher accuracy in diagnosing VZV than using the IgA, IgG, or IgM detection separately.

2.5. Equivocal Sample Analysis Shows Higher Sensitivity/Accuracy for CLIA Than ELISA in Combined Antibody Detection

To assess the diagnostic ability of the CLIA kit to diagnose the samples classified as equivocal by a standard ELISA tests, individual and combined detection of IgA, IgG, and IgM were conducted with the CLIA-based diagnostic approach in the eight-patient cohort. These results were analyzed along with previous cohort results and compared to those from ELISA. The comparative analysis was based on previously obtained IgA, IgG, and IgM criteria (>78662 RLU, >23450 RLU, and >89634 RLU, respectively) (Figure 2A–C) and the validation criteria of standard ELISAs.

For IgA detection, only patient samples 24 and 20 were diagnosed as positive by ELISA and CLIA, respectively (Figure 1, Table 2). No significant difference was observed in IgA levels between equivocal and negative controls (p-value = 0.567) (Figure 3A). Regarding IgG detection, all of these eight patients were determined equivocal, and none was positive by ELISA (Figure 1, Table 2). However, CLIA-based diagnostic revealed that two patient samples (patients 13 and one sample from 24) were positive (RLU > 23450), and the analysis of all the eight patient samples displayed RLU values significantly different from that of the negative controls (p-value < 0.001) (Table 2 and Figure 3B). Regarding IgM assessment, most of the ELISA-based equivocal samples were diagnosed as positive by CLIA (Figure 3C, Table 2). Overall, the combined diagnostic kits could detect lower immunoglobulin concentrations compared to ELISAs. These results suggest that CLIA-based diagnostic kits perform better than ELISAs in diagnosing varicella–zoster, especially in combination (Tables 1 and 2).

Table 2. VZV-gE-specific IgG, IgA, and IgM diagnostic results in the eight equivocal patient results.

Pat.	n°	RLU (OD$_{450}$) Values			CLIA (ELISA) Results			Agreement Positivity (IgA or IgG or IgM)
		IgA	IgG	IgM	IgA	IgG	IgM	
6	1	34000 (0.099)	14883 (0.131)	795717 (0.09)	N (N)	N (E)	P (N)	
	2	19610 (0.076)	18775 (0.076)	932321 (0.076)	N (N)	N (N)	P (N)	P (E)
	3	14062 (0.089)	8502 (0.103)	1011533 (0.083)	N (N)	N (E)	P (N)	
11	1	2042 (0.136)	15398 (0.182)	717862 (0.073)	N (E)	N (E)	P (N)	
	2	8306 (0.089)	18254 (0.188)	816205 (0.088)	N (N)	N (E)	P (N)	P (E)
13	1	39269 (0.198)	23450 (0.159)	221901 (0.192)	N (E)	P (E)	P (E)	
	2	20987 (0.097)	47135 (0.097)	873343 (0.187)	N (N)	P (N)	P (E)	P (E)
	3	33650 (0.147)	27064 (0.129)	415041 (0.191)	N (E)	P (E)	P (N)	
15	1	69084 (0.157)	18366 (0.191)	995029 (0.169)	N (E)	N (E)	P (E)	
	2	34439 (0.069)	13626 (0.187)	854559 (0.188)	N (N)	N (E)	P (E)	P (E)
	3	56329 (0.11)	12408 (0.185)	1017070 (0.192)	N (E)	N (E)	P (E)	
18	1	34252 (0.071)	8940 (0.091)	64208 (0.111)	N (N)	N (N)	N (E)	
	2	26172 (0.091)	10172 (0.188)	78205 (0.191)	N (N)	N (E)	N (E)	N (E)
20	1	102290 (0.081)	7684 (0.198)	103009 (0.236)	P (N)	N (N)	P (P)	
	2	102454 (0.069)	8749 (0.099)	94009 (0.068)	P (N)	N (E)	P (N)	P (P)
	3	187560 (0.146)	11790 (0.151)	100154 0.197)	P (E)	N (E)	P (E)	
21	1	25117 (0.198)	3836 (0.075)	30841 (0.062)	N (E)	N (N)	N (N)	
	2	78099 (0.084)	15208 (0.153)	88775 (0.074)	N (N)	N (E)	N (N)	N (E)
24	1	3365 (0.513)	27405 (0.189)	571206 (0.401)	N (P)	P (E)	P (P)	
	2	69084 (0.601)	13048 (0.089)	605542 (0.285)	N (P)	N (N)	P (P)	P (P)
Total		-	-		3(2)/20	4(1)/20	16(3)/20	6(2)/8

E: Equivocal; P: Positive; N: Negative; RLU: Relative Light Unit. The number of samples for each patient is determined by grey difference.

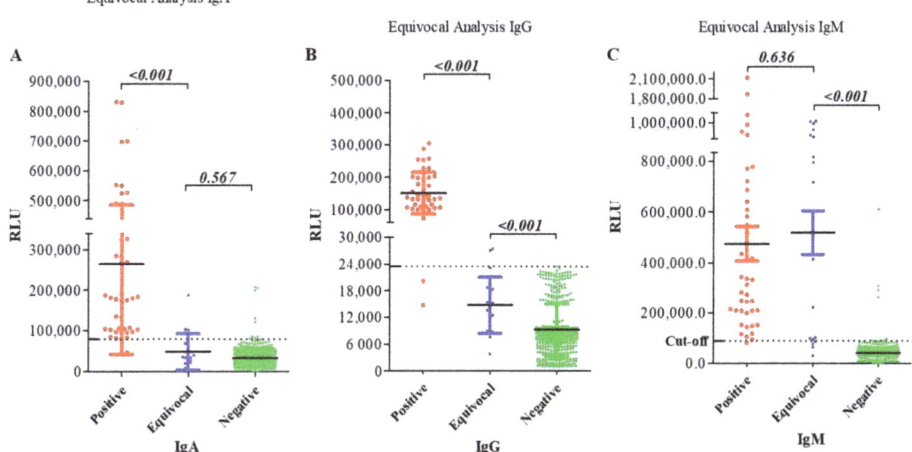

Figure 3. VZV-gE specific IgA, IgG, and IgM detection results and antibody levels in the patient cohort. Analysis of specific VZV serum antibody levels in highly positive (42 samples from 21 patients) and equivocal (20 samples from eight patients) patients revealed different levels in IgA (**A**), IgG (**B**), and IgM (**C**) antibody titers, as defined by the automated relative light units (RLU). The black bars in each distribution represent the mean, respectively, associated with the standards error of means (SEM). The dotted line indicates the cut-off values (>78,662 for IgA (**A**), >23,450 for IgG (**B**), and >89,634 for IgM (**B**)). RLU: relative light unit.

2.6. Diagnostic Performance of CLIA-Based Immunoglobulin Diagnosis Regarding Equivocal Patients

We aimed to determine the diagnostic performance of the CLIA-based IgA, IgG, and IgM detection kit in diagnosing VZV infection in a random population. To this end, we considered the equivocal samples as positive and used the whole cohort, which consisted of the 62 independent samples from included patients and the 453 negative samples from healthy donors. The ROC analysis showed sensitivities of 74.2% (IC95%: 61.5–84.5%), 69.4% (IC95%; 56.3–80.4%), and 93.6% (IC95%: 84.3–98.2%), and specificities of 96.0% (IC95%: 93.8–97.6%), 99.8% (IC95%: 98.0–100%), and 98.0% (IC95%: 96.3–99.1%) for the diagnostic of VZV-gE IgA, IgG, and IgM, respectively (Figure 2D–F, Table 1). Interestingly, when the three VZV-gE IgA, IgG, and IgM detections were combined, the diagnostic performance was enhanced to sensitivity, specificity, and an overall agreement of 98.4%, 100%, and 99.8%, respectively (Table 1). Altogether, these analyses confirm that CLIA-based IgM detection alone or combined with IgA and IgG detection provides better diagnostic accuracy in diagnosing varicella and herpes zoster.

2.7. Antibody Titer Analysis in VZV Patients Suggests a Primary Infection in Equivocal Patients

Patient clinical data along with CLIA-based diagnostic results were analyzed. As a result, IgM titer was higher in all patients, while IgA and IgG titers were lower in patients with equivocal results (Figures 3 and 4A). Moreover, since none of the included patients declared to have received VZV vaccine and with low IgA/IgG and high IgM titer, it was concluded that patients with equivocal results experienced varicella infection (primary infection). The high IgM detection associated with the presence of typical VZV-associated symptoms as early as two days in these patients (Supplementary Table S1) confirmed an acute phase of the varicella infection. However, unlike these eight patients, the 21 true-positive patients with high IgA, IgG, and IgM titers experienced an acute phase of viral reactivation (herpes zoster) or viral reinfection.

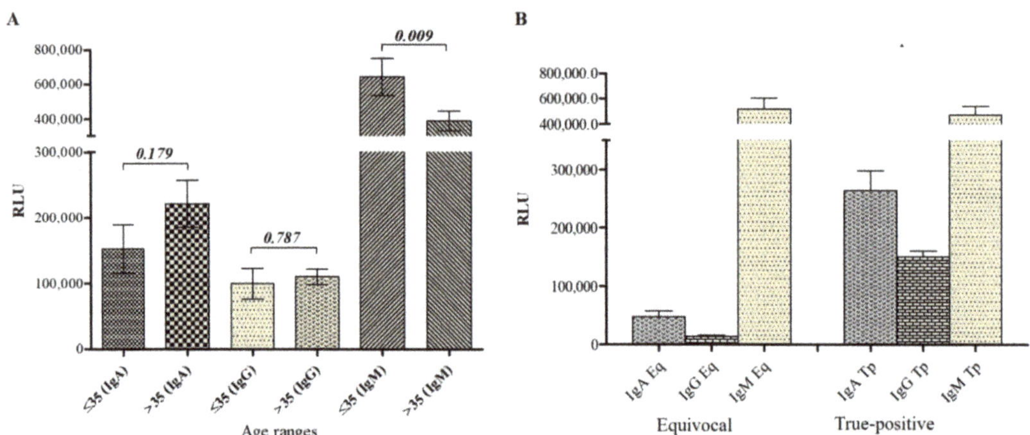

Figure 4. Antibody titer analysis in VZV patients. (**A**). Analysis of CLIA-based diagnostic results demonstrated that patients with equivocal diagnoses were in the acute primary infection state. The high level of IgM in these patient samples corresponds to the early production of adaptative immunity (IgM), and the low titer of IgG corresponds to the gradient production of memory immunity. In contrast, patients with a high level of IgG and IgM were probably in the acute state of either reactivation or reinfection-associated herpes zoster. The high level of IgG and IgM demonstrate a simultaneous presence of active/acute memory immunity. Eq: equivocal; Tp: True-positive. (**B**). Variation of serum antibody level with age. Result analysis of IgA, IgG, and IgM antibody levels regarding the age range revealed a difference in IgM level (*p*-value < 0.05). RLU: relative light unit.

2.8. VZV-gE-Specific IgM Titer Negatively Correlates with Age

To assess whether there is any significant difference between the IgA, IgG, and IgM levels regarding VZV-patient age, the patient cohort was divided into two groups ($\leq$35 years old and >35 years old). CLIA-associated diagnostic results from the 28 patients with reported age were considered (Supplementary Table S1). Analyses revealed that while no difference was observed in IgA/IgG titer, there was a significant difference in IgM level between the two groups. IgM level was significantly higher in young adults (below 35 years old) than in elder (Figure 4B).

3. Discussion

Reliable assays in the immunological diagnosis of VZV with the best performance are essential in the era of vaccine program implementation, their effectiveness evaluation, and for VZV-associated disease monitoring [12,14,18–20,23]. Most laboratories currently diagnose VZV-associated diseases from patient clinical symptoms, which is biased and may result in misdiagnoses identifying other (herpesvirus) infections with similar symptoms, such as HSV [25,37]. Moreover, VZV infections may usually present in atypical forms [38]. Various VZV immunodiagnostic tests developed so far lack sensitivities, and the most sensitive are unevenly available worldwide [9,13,23–27]. Here, we used a purified VZV-gE protein to develop and validate a highly-sensitive/accurate and automated CLIA approach for detecting IgA, IgG, and IgM specific to VZV in patient blood.

The developed CLIA diagnostic approach detected antibodies in VZV-patient samples with high accuracy/specificity and proportionally to the titer. Interestingly, this approach could detect very low antibody titers and determine positivity in most ELISA-based equivocal results. Moreover, the testing process was simple, fully automated, and thus suitable for high-throughput screening situations. The highly accurate results could be obtained as short as 50 min, with enhanced performance, making this CLIA-based diagnostic a better VZV-diagnostic tool than ELISA. The following Figure 5 describes the principle of CLIA approach as applied in Kaeser automate (Kangrun Biotech, Guangzhou, China).

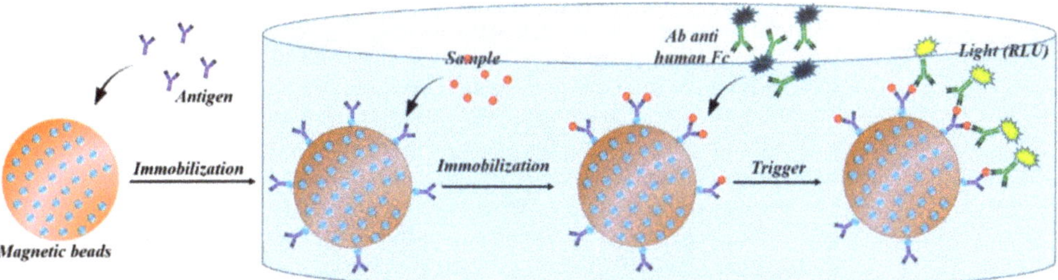

Figure 5. CLIA-based diagnostic assay principle. Purified VZV-gE antigen is immobilized onto metal beads and saturated with BSA. In the machine, a small amount of controls or test samples are added to the test tube and incubated. The test tube is washed to remove any unbound human immunoglobulin (h-Ig). A pre-labeled anti-human Ig conjugate is added to the test tubes. Then, a prepared substrate is added and catalyzed by the pre-labeled enzyme to produce a fluorescence, which is directly proportional to the amount of human anti-antigen Ig captured on the beads.

The high accuracy demonstrated by the developed diagnostic kit in VZV-specific IgA, IgG, and IgM detection in blood samples was expected. Practically, infection with VZV induces robust antibody response, including IgA, IgG, and IgM antibodies produced mainly against VZV-gE and VZV-gI [1,39,40]. In fact, during the viral replication cycle, the VZV-gE is the most abundant glycoprotein produced and expressed on the VZV-infected cell surface, thus playing a central role in anti-VZV antibody production [1,39,40]. Moreover, while IgM is responsible for rapid and early immunity, long-term humoral immunity is initiated by the production of high-affinity IgG or IgA antibody. These circulating antibodies, especially the VZV-gE specific antibodies, are known to have the highest affinity to and neutralizing effect against VZV infection. During acute infections, IgA, IgG, and IgM antibody production is higher. However, in the absence of symptoms and in the earlier stage of infection, immunoglobulin production is low, and the most serum immunoglobulins produced are specifically directed toward VZV-gE, supporting our choice to use VZV-gE protein as serological antigen for developing this highly specific/accurate VZV diagnostic test [41,42]. Interestingly, Anna Grahn et al. [43] demonstrated that the use of VZV-gE in the detection of intrathecal specific antibodies is highly specific, without HSV non-specific reaction. Additionally, testing IgA together with IgG and IgM is crucial and has an added value in VZV diagnosis, because secretory IgA are mainly produced during VZV infection, as mucosal epithelial cells that mainly secrete IgA are the first cells to be infected with VZV.

The performance of some current commercially available VZV immunodiagnostic tests, such as VZV TRFIA and VaccZyme™ EIA, has been evaluated and reported [27]. For instance, from unvaccinated healthcare workers, VaccZyme™ EIA shows IgG detection sensitivities of up to 54.2% and specificities above 98.6%. On a comparable unvaccinated cohort, our developed CLIA-based VZV-gE IgG detection test showed better performance, with diagnostic sensitivity and specificity of 95.2% (IC95%: 83.8–99.4%) and 100% (IC95%: 99.2–100%), respectively excluding equivocal patients, and of 69.4% (IC95%: 56.3–80.4%) and 99.8% (IC95%; 98.0–100%), respectively considering equivocal patients (Table 1 and Figure 2).

There are a few available diagnostic kits that combine the simultaneous detection of VZV-IgA, IgG, and IgM. Moreover, the diagnostic performance of available IgM ELISA tests is lacking, especially in unvaccinated people [40,44,45]. Our CLIA-based detection kit, with high sensitivities and specificities in combined detection of IgA, IgG, and IgM specific to VZV in patient's blood (Figure 2, tables 1 and 2), would be beneficial in routine diagnosis of varicella and herpes zoster. Furthermore, it can be of added value for post-vaccination immunity assessment, as good performance is expected in detecting low antibody levels commonly faced in vaccine recipients.

IgM is produced in high titer during the acute phase of primary infection [37,40], while IgG and IgA titers are low. Assessing IgA/IgG in this infection stage may result in false-negative-to-equivocal results. In the context of VZV infection diagnosis, such as in this study, such situations are usual [12] and require diagnosis confirmation 7 to 14 days later. However, in the presence of VZV infection symptoms such as rashes, the detection of VZV-specific IgM confirms the acute phase of the infection, although without specifying between primary, self-infection or reinfection, and viral reactivation. In the ELISA-based equivocal patients, the CLIA-based diagnostic approach showed low IgA/IgG and high IgM levels, suggesting that these patients (especially 6/8 patients) suffered from varicella in the acute phase. Pertinently, patients with equivocal results (except patients 18 and 21) visited the hospital as early as 2 to 3 days after the symptom onsets (Supplementary Table S1), thus supporting the hypothesis of an acute primary infection. For instance, studies of experimental simian varicella virus infection in monkeys demonstrated that IgG appears five days after IgM production, decreasing without competing with IgG [46]. A contrario, patients with a high level of IgA/IgG/IgM probably experienced herpes zoster.

The use of PCR in VZV diagnosis is preferred and recommended, as it is the most sensitive method to confirm varicella-zoster infection in vesicular lesions or scabs [12,24,28]. However, in the absence of rashes, this method is limited, with a decreased sensitivity, and results in false-negative when other samples, including blood, saliva, and cerebrospinal fluid (CSF), are used. Interestingly, it has been reported that in the absence of rashes, the use of blood and CSF to detect VZV-specific IgG by immunological methods yields more sensitivity than PCR for DNA detection [29,30]. Moreover, similar to other commercial immunological tests, PCR-based diagnostics are not widely available [12], and it is expensive and leads to patient compliance [47]. Altogether, our CLIA-based diagnostic tests filled these gaps and would be helpful in public health laboratories for routine varicella-zoster disease diagnosis, control outbreak situations [12], and varicella-zoster seroepidemiological studies for vaccine implementation purposes [14,18–20,23].

In this cohort, analysis of IgA, IgG, and IgM levels showed no significant antibody variation regarding gender (data not shown). However, regarding the age, it came out that while no difference in IgA and IgG level was found, the IgM level showed differences between young adults (below 35) and older. IgM tended to be higher in the youngest than in the eldest, which does not corroborate other studies, in which the antibody titers were proportional with age [20,48]. A larger population size would be preferred to draw a better relative conclusion. However, the conclusion on features and performance of our developed tests remains unaffected and valid.

However, although the population size permitted to validate the diagnostic approach, it was not large enough to better evaluate the performance of the test on a representative population scale, precisely to determine the predictive positive and negative values. For the same reason, evaluation of the correlation between each antibody and age range could have been biased as well. Therefore, future investigation with a large and representative population (including vaccinated and non-vaccinated) will better evaluate the performance of this diagnostic approach and study the immune response regarding age range. Moreover, it is suggested to use other samples, including saliva, which is thought to contain higher concentration of antibodies, specifically IgA in VZV infection, for a conclusive added value of IgA detection in the conventional immunodiagnostic kit.

In conclusion, detecting VZV-gE-specific IgA, IgG, and IgM using the developed kits based on the CLIA approach provided high sensitivity/accuracy and a rapid practical method for diagnosing VZV in unvaccinated individuals or determining VZV immune status after natural infection. This approach is simple, does not require outstanding trainees, and is suitable in high-throughput diagnosis situations.

4. Materials and Methods

4.1. Patient and Clinical Samples

This study was carried out under the approval (n° 2021-ky269) of the Medical Ethics Committee of the First Affiliated Hospital of the University of Science and Technology of China (USTC). From June to December 2020, a flow of patients was received in the hospital dermatological department for rashes, pimples, and other skin issues. Based on the presence of typical VZV symptoms (including paresthesia, localized pains, pimples, and non-oral and genital rashes), several patients were diagnosed as VZV-infected patients and included after obtaining free consent of participation. Two to three blood samples were collected into EDTA and dry tubes from each enrollee to investigate immunoglobulin (Ig) A, IgG, and IgM in plasma and serum, respectively. ELISA ab108781 (IgA), ab108782 (IgG), and ab108783 (IgM) tests (Abcam) were used as standards to exclude patient samples with negative results for all IgA, IgG, and IgM and retained those with at least one positive/equivocal result for IgA, IgG, or IgM, and thus stratified as true-positive or equivocal group. A total of 29 patients were retained, from which 62 blood samples were obtained for testing the developed CLIA diagnostic approach. Clinical patient data were obtained and listed in Supplementary Table S1.

Negative control samples were collected to assess the diagnostic accuracy. This cohort contained 453 samples from random healthy consenting people who did not report having suffered from or having been diagnosed positive for VZV infections and did not receive any VZV vaccines. All plasmas and sera were retrieved from EDTA (using Ficoll; density: 1.077) and dry tubes, respectively, by centrifugation. Retrieved plasmas/sera were treated with 1% TNBP and 1% Triton X-100 to completely denature any potential viruses [49] and stored at −20 °C (or −80 °C) until use.

4.2. Enzyme-Linked Immunosorbent Assay Tests

As aforementioned, three 96-well plate ELISA tests (Abcam ab108781, ab108782, and ab108783) were used to detect VZV-specific antibodies (IgA/IgG/IgM) in the enrolled patient samples as a complementary confirmation step. The ELISA tests were performed following the manufacturer's instructions [15]. Before testing, samples were first virus-inactivated and then diluted accordingly with dilution buffer (PBS). For testing sera/plasmas, the manufacturer's instructions were followed. The resulting yellow color intensity was measured at OD450 using a microplate reader. Each ELISA test was triplicated, and the data was graphed using GraphPad Prism 5 software.

Patient diagnostic results were determined from OD450 at dilution 1/100, as mentioned in the diagnostic kit leaflet. As a result of considerable background, the OD of the blank (PBS) was deducted from OD450 values of each sample result. Thus, a patient was considered positive when OD450 > 0.2, equivocal when OD450 was between 0.1 and 0.2, and negative when OD450 < 0.1. All patients negative for IgA, IgG, and IgM were systematically excluded from the study, while the retained patients were divided as true-positive or equivocal.

4.3. VZV Glycoprotein E Antigen Preparation

To develop the highly-sensitive diagnostic approach detecting VZV-specific IgA, IgG, and IgM in VZV-patient blood samples, the surface antigen glycoprotein E of VZV (VZV-gE) was produced from insect cell cultures using the baculovirus-based vector expression system (BVES) (Invitrogen-ThermoFisher Scientific, Waltham, MA, USA).

4.3.1. Cell Cultures

Two insect cell lines, including *Spodoptera frugiperda* (*Sf9*) and *Trichoplusia ni* (*High Five, Hi5*), were used to produce the baculoviruses carrying the gene of interest and express gE protein, respectively. *Sf9* and *Hi5* were cultured at 27 °C in SIM-SF and SIM-HF medium (Sino Biological Inc., Beijing, China), respectively, and supplemented with 10% (*v/v*) heat-inactivated fetal bovine serum (FBS) and 1× penicillin/streptomycin. *Sf9* cell lines were maintained in both adherent and suspension cultures, while *Hi5* cell lines were only cultured in suspension culture. Adherent cell culture was carried out in 6 cm TC plates, and the suspension cultures were maintained in sterile autoclaved Erlenmeyer flasks in a 110 rpm spin shaker (27 °C). The suspension *Hi5* cell culture was diluted to 0.7 to 1 million cells every two based on cell viability and density. The adherents *Sf9* cell cultures were detached and diluted every three days based on cell viability and confluence. The cell viability was assessed under a fluorescence microscope using Trypan blue dye (0.4%) and counted using a hemocytometer.

4.3.2. Molecular Cloning, Expression, and Purification of VZV-gE Protein

Briefly, the sequence of the mature extracellular region of VZV-gE (GenBank Accession number MH709377.1) was retrieved by PCR from the General Biosystem company's synthetic construct using the following forward and reverse primers: 5′-ATTTCCAAGGTTCTT CCGTCTTGCGATACGATGATTTTCACATC-3′ and 5′-GACAAGCTTGGTACTTAATATCG TAGAAGTGGTGACGTTCCGGG-3′, respectively. In the meantime, the His-tag-modified transfer vector (pI-SUMO-Star-His) was linearized using primers (forward: 5′CTTCTACGA TATTAAGTACCAAGCTTGTCGAGAAGTACTAGAGG3′ and reverse: 5′TATCGCAAGAC GGAAGAACCTTGGAAATAAAGATTCTCGCTGCC3′) containing sequences that overlap the VZV-gE 5′ and 3′ end sequences. The linear fragments were ligated following the Gibson Assembly method's instructions. The successful construct was transposed into bacmid using *DH10Bac E. coli* strain and purified for transfecting *Spodoptera frugiperda* (*Sf9*) insect cell lines, which produced recombinant VZV-gE baculoviruses. Expanded recombinant baculovirus stock was used to infect two million *Trichoplusia ni* (*High Five, Hi5*) insect cell lines for expressing the recombinant VZV-gE protein. The protein was harvested three days post-infection by high-speed centrifugation and purified from the supernatant.

A couple of purifications steps, including membrane diafiltration (Vivaflow 200), dialysis, ion-nickel column purification, size-exclusion chromatographic purification, and ultra-centrifugation were conducted to purify the protein. The purified protein (Supplementary Figure S3) was stored in HEPES buffer saline (HBS: 20 mM HEPES, 250 mM NaCl), an amine-free buffer, which is required for the further experiments.

4.4. Preparation and Validation of the CLIA-Based Diagnostic Kit

The highly purified VZV-gE protein was employed to make the CLIA-based diagnostic kit. The purified VZV-gE protein was first biotinylated using EZ-Link Sulfo-NHS-LC-LC-Biotin, No-Weigh™ Format kit (Thermo Fisher, n°A35358, Waltham, MA, USA) following the manufacturer's instructions. Then, the biotinylated protein was immobilized onto magnetic beads using an Invitrogen Dynabeads™ MyOne™ Streptavidin C1 kit (Thermo Fisher), following the manufacturer's instructions, and further blocked (or saturated) with 2% of bovine serum albumin (BSA) to avoid non-specific interactions or background. Immobilizing the antigen protein onto a solid phase (here beads) is necessary for immunoblotting IgA, IgG, and IgM antibodies on a solid phase. The detection procedure below was performed with a fully automatic chemical luminescent immune analyzer, Kaeser 1000 (Kangrun Biotech, Guangzhou, China). Secondary antibodies anti-human IgA, IgG, or IgM conjugated with acridinium were used to detect the caught VZV-gE specific IgG, IgA, or IgM antibodies, respectively. The detected chemiluminescent signal over the background signal was automatically obtained as relative light units (RLU).

These collections, which contain all the described buffers and components for CLIA of VZV gE-specific IgA, IgG, and IgM, are referred to as VZV-gE-IgA, VZV-IgG, and VZV-gE-IgM kits here. As described above, each diagnostic kit was developed independently, with the corresponding secondary antibody conjugated with acridinium.

A first test batch of a two-fold serial dilution of the ELISA-based true-positive and healthy plasmas/sera was conducted to assess the reliability of the antibody detection kit regarding sample dilution. A subsequent CLIA test was performed to determine the diagnostic kit performances.

4.5. Statistical Analysis

ELISA tests were triplicate, and the results were transformed, fitted, and presented as mean $\pm$ SD. To determine the optimal cut-off values (criteria) and evaluate the diagnostic characteristics of VZV-gE-IgA, IgG, and IgM kits, receiver-operating characteristic (ROC) analyses were performed using MedCalc software. Thus, the specificity and sensitivity of the gE-specific IgA, IgG, and IgM detection kits were determined according to the following formulas:

- Sensitivity (%) = 100 $\times$ [True Positive/(True Positive + False Negative)];
- Specificity (%) = 100 $\times$ [True Negative/(True Negative + False Positive)];
- Overall agreement (%) = (True Positive + True Negative)/Total Tests.

A Mann–Whitney test was used to assess any significant variation of VZV gE-specific IgA, IgG, or IgM level between equivocal and true-positive ELISA-based categories. The same analysis was used to assess any significant correlation of the antibody levels regarding the age ranges. An analysis of variance (ANOVA) test was conducted using the Kruskal–Wallis approach to determine any difference of antibody level between the three independent groups, including positive, equivocal, and negative. A p-value less than 0.05 defined a hypothesis as statistically significant. All the above analyses were integrated into GraphPad Prism5.

Supplementary Materials: The following are available online at https://www.mdpi.com/article/10.3390/pathogens11010066/s1. Supplementary Figure S1: Flow chart of patient inclusion and analyses, Supplementary Table S1: Epidemiological and clinical patient data, Supplementary Figure S2: ELISA results of all the 29 included patients, Supplementary Figure S3: Purification of VZV-gE recombinant protein using baculovirus-based vector expression system (BVES), Supplementary Figure S4: Validation of VZV-gE specific IgA, IgG, and IgM detection based on CLIA approach.

Author Contributions: Conceptualization, A.J.K.K. and T.J.; data curation, A.J.K.K.; formal analysis, A.J.K.K.; funding acquisition, T.J.; investigation, S.Z.; methodology, A.J.K.K.; project administration,

S.Z. and T.J.; resources, A.J.K.K. and T.J.; software, A.J.K.K.; supervision, S.Z. and T.J.; validation, A.J.K.K., S.Z. and T.J.; writing—original draft, A.J.K.K.; writing—review and editing, A.J.K.K., S.Z., J.X., A.Z., H.M., G.X., Y.D., F.A.N.B., A.M., Z.D., Y.Y., C.F., W.Z., R.C., K.Z. and T.J. All authors have read and agreed to the published version of the manuscript.

Funding: This research was funded by the Strategic Priority Research Program of the Chinese Academy of Sciences (Grant No. XDB29030104), the National Natural Science Fund (Grant No.: 31870731 and 31971129), the Fundamental Research Funds for the Central Universities, Jack Ma Foundation, and the 100 Talents Program of The Chinese Academy of Sciences, the Chinese Government scholarship program, and the CAS-TWAS scholarship program.

Institutional Review Board Statement: The study was conducted according to the guidelines of the Declaration of Helsinki and approved by the Ethics Committee of the First Affiliated Hospital of the University of Science and Technology of China (USTC), n° 2021-ky269.

Informed Consent Statement: Informed consent was obtained from all subjects involved in the study. Written informed consent has been obtained from the patients to publish this paper.

Data Availability Statement: Restrictions apply to the availability of the patients data. Patient data was obtained from the First Affiliated Hospital of the University of Science and Technology of China (USTC) and are available from the corresponding authors with the permission of the First Affiliated Hospital of the University of Science and Technology of China (USTC), while patient sample-related experimental research data was obtained from our Laboratory.

Acknowledgments: We would like to thank the staff and patients at the Department of Dermatology and the Anhui Provincial Hospital, The First Affiliated Hospital of USTC, for their support in providing samples and collecting clinical data.

Conflicts of Interest: Tengchuan Jin, Arnaud John KOMBE KOMBE, and Huan Ma in USTC have jointly applied for a joint patent related to the antibody detecting kits. Other authors declare no conflicts of interest.

References

1. Arvin, A.M. Varicella-zoster virus. *Clin. Microbiol. Rev.* **1996**, *9*, 361–381. [CrossRef]
2. Baird, N.L.; Zhu, S.; Pearce, C.M.; Viejo-Borbolla, A. Current In Vitro Models to Study Varicella Zoster Virus Latency and Reactivation. *Viruses* **2019**, *11*, 103. [CrossRef]
3. Gershon, A.A.; Breuer, J.; Cohen, J.I.; Cohrs, R.J.; Gershon, M.D.; Gilden, D.; Grose, C.; Hambleton, S.; Kennedy, P.G.; Oxman, M.N.; et al. Varicella zoster virus infection. *Nat. Rev. Dis. Primers* **2015**, *1*, 15016. [CrossRef]
4. Kennedy, P.G.E.; Gershon, A.A. Clinical Features of Varicella-Zoster Virus Infection. *Viruses* **2018**, *10*, 609. [CrossRef]
5. Abdullahi, A.M.; Sarmast, S.T.; Jahan, N. Viral Infections of the Central Nervous System in Children: A Systematic Review. *Cureus* **2020**, *12*, e11174. [CrossRef]
6. Zhou, J.; Li, J.; Ma, L.; Cao, S. Zoster sine herpete: A review. *Korean J. Pain* **2020**, *33*, 208–215. [CrossRef]
7. Nagel, M.A.; Gilden, D. Neurological complications of varicella zoster virus reactivation. *Curr. Opin. Neurol.* **2014**, *27*, 356–360. [CrossRef]
8. Nichols, R.A.; Averbeck, K.T.; Poulsen, A.G.; Al Bassam, M.M.; Cabral, F.; Aaby, P.; Breuer, J. Household size is critical to varicella-zoster virus transmission in the tropics despite lower viral infectivity. *Epidemics* **2011**, *3*, 12–18. [CrossRef]
9. Heywood, A.E.; Macartney, K.K.; MacIntyre, C.R.; McIntyre, P.B. Current developments in varicella-zoster virus disease prevention. A report on the varicella-zoster virus workshop convened by the National Centre for Immunisation Research and Surveillance of Vaccine Preventable Diseases on 16–17 November 2006. *Commun. Dis. Intell. Q. Rep.* **2007**, *31*, 303–310.
10. Bonmarin, I.; Santa-Olalla, P.; Levy-Bruhl, D. Modelling the impact of vaccination on the epidemiology of varicella zoster virus. *Rev. Epidemiol. Sante Publique* **2008**, *56*, 323–331. [CrossRef]
11. De Donno, A.; Kuhdari, P.; Guido, M.; Rota, M.C.; Bella, A.; Brignole, G.; Lupi, S.; Idolo, A.; Stefanati, A.; Del Manso, M.; et al. Has VZV epidemiology changed in Italy? Results of a seroprevalence study. *Hum. Vaccin. Immunother.* **2017**, *13*, 385–390. [CrossRef] [PubMed]
12. Leung, J.; Harpaz, R.; Baughman, A.L.; Heath, K.; Loparev, V.; Vazquez, M.; Watson, B.M.; Schmid, D.S. Evaluation of Laboratory Methods for Diagnosis of Varicella. *Clin. Infect. Dis.* **2010**, *51*, 23–32. [CrossRef]
13. Adriana Lopez, J.L.; Scott Schmid, M.M. *Manual for the Surveillance of Vaccine-Preventable Diseases: Chapter 7—Varicella*; Centers for Disease Control and Prevention: Atlanta, GA, USA, 2018.
14. Senders, S.D.; Bundick, N.D.; Li, J.; Zecca, C.; Helmond, F.A. Evaluation of immunogenicity and safety of VARIVAX New Seed Process (NSP) in children. *Hum. Vaccin. Immunother.* **2018**, *14*, 442–449. [CrossRef]

15. Sauerbrei, A.; Wutzler, P. Serological detection of varicella-zoster virus-specific immunoglobulin G by an enzyme-linked immunosorbent assay using glycoprotein antigen. *J. Clin. Microbiol.* **2006**, *44*, 3094–3097. [CrossRef]

16. Edmunds, W.J.; Brisson, M. The effect of vaccination on the epidemiology of varicella zoster virus. *J. Infect.* **2002**, *44*, 211–219. [CrossRef]

17. Hope-Simpson, R.E. The Nature of Herpes Zoster: A Long-Term Study and a New Hypothesis. *Proc. R. Soc. Med.* **1965**, *58*, 9–20. [CrossRef] [PubMed]

18. Wu, Q.S.; Wang, X.; Liu, J.Y.; Chen, Y.F.; Zhou, Q.; Wang, Y.; Sha, J.D.; Xuan, Z.L.; Zhang, L.W.; Yan, L.; et al. Varicella outbreak trends in school settings during the voluntary single-dose vaccine era from 2006 to 2017 in Shanghai, China. *Int. J. Infect. Dis.* **2019**, *89*, 72–78. [CrossRef]

19. Yue, C.; Li, Y.; Wang, Y.; Liu, Y.; Cao, L.; Zhu, X.; Martin, K.; Wang, H.; An, Z. The varicella vaccination pattern among children under 5 years old in selected areas in china. *Oncotarget* **2017**, *8*, 45612–45618. [CrossRef]

20. Luan, L.; Shen, X.; Qiu, J.; Jing, Y.; Zhang, J.; Wang, J.; Zhang, J.; Dong, C. Seroprevalence and molecular characteristics of varicella-zoster virus infection in Chinese children. *BMC Infect. Dis.* **2019**, *19*, 643. [CrossRef] [PubMed]

21. Haugnes, H.; Flem, E.; Wisloff, T. Healthcare costs associated with varicella and herpes zoster in Norway. *Vaccine* **2019**, *37*, 3779–3784. [CrossRef]

22. Marin, M.; Guris, D.; Chaves, S.S.; Schmid, S.; Seward, J.F.; Advisory Committee on Immunization Practices, Centers for Disease Control and Prevention (CDC). Prevention of varicella: Recommendations of the Advisory Committee on Immunization Practices (ACIP). *MMWR Recomm. Rep.* **2007**, *56*, 1–40.

23. Provost, P.J.; Krah, D.L.; Kuter, B.J.; Morton, D.H.; Schofield, T.L.; Wasmuth, E.H.; White, C.J.; Miller, W.J.; Ellis, R.W. Antibody assays suitable for assessing immune responses to live varicella vaccine. *Vaccine* **1991**, *9*, 111–116. [CrossRef]

24. Nikolic, D.; Kohn, D.; Yen-Lieberman, B.; Procop, G.W. Detection of Herpes Simplex Virus and Varicella-Zoster Virus by Traditional and Multiplex Molecular Methods. *Am. J. Clin. Pathol.* **2019**, *151*, 122–126. [CrossRef] [PubMed]

25. Wilson, D.A.; Yen-Lieberman, B.; Schindler, S.; Asamoto, K.; Schold, J.D.; Procop, G.W. Should varicella-zoster virus culture be eliminated? A comparison of direct immunofluorescence antigen detection, culture, and PCR, with a historical review. *J. Clin. Microbiol.* **2012**, *50*, 4120–4122. [CrossRef]

26. Yamamoto, T.; Aoyama, Y. Detection of multinucleated giant cells in differentiated keratinocytes with herpes simplex virus and varicella zoster virus infections by modified Tzanck smear method. *J. Dermatol.* **2021**, *48*, 21–27. [CrossRef]

27. Maple, P.A.; Breuer, J.; Quinlivan, M.; Kafatos, G.; Brown, K.E. Comparison of a commercial Varicella Zoster glycoprotein IgG enzyme immunoassay with a reference time resolved fluorescence immunoassay (VZV TRFIA) for measuring VZV IgG in sera from pregnant women, sera sent for confirmatory testing and pre and post vOka vaccination sera from healthcare workers. *J. Clin. Virol.* **2012**, *53*, 201–207. [CrossRef] [PubMed]

28. Schmutzhard, J.; Merete Riedel, H.; Zweygberg Wirgart, B.; Grillner, L. Detection of herpes simplex virus type 1, herpes simplex virus type 2 and varicella-zoster virus in skin lesions. Comparison of real-time PCR, nested PCR and virus isolation. *J. Clin. Virol.* **2004**, *29*, 120–126. [CrossRef]

29. Kennedy, P.G. Issues in the Treatment of Neurological Conditions Caused by Reactivation of Varicella Zoster Virus (VZV). *Neurotherapeutics* **2016**, *13*, 509–513. [CrossRef] [PubMed]

30. Nagel, M.A.; Gilden, D. Update on varicella zoster virus vasculopathy. *Curr. Infect. Dis. Rep.* **2014**, *16*, 407. [CrossRef]

31. Majumder, P.; Shetty, A.K. Comparison between ELISA and chemiluminescence immunoassay for the detection of Hepatitis C virus antibody. *Indian J. Microbiol. Res.* **2017**, *4*, 353–357. [CrossRef]

32. Chen, D.; Zhang, Y.; Xu, Y.; Shen, T.; Cheng, G.; Huang, B.; Ruan, X.; Wang, C. Comparison of chemiluminescence immunoassay, enzyme-linked immunosorbent assay and passive agglutination for diagnosis of Mycoplasma pneumoniae infection. *Ther. Clin. Risk Manag.* **2018**, *14*, 1091–1097. [CrossRef]

33. de Ory, F.; Minguito, T.; Balfagon, P.; Sanz, J.C. Comparison of chemiluminescent immunoassay and ELISA for measles IgG and IgM. *APMIS* **2015**, *123*, 648–651. [CrossRef] [PubMed]

34. Ma, H.; Zeng, W.; He, H.; Zhao, D.; Yang, Y.; Jiang, D.; Qi, P.Y.; He, W.; Zhao, C.; Yi, R.; et al. COVID-19 diagnosis and study of serum SARS-CoV-2 specific IgA, IgM and IgG by chemiluminescence immunoanalysis. *medRxiv* **2020**. Available online: https://www.medrxiv.org/content/10.1101/2020.04.17.20064907v2 (accessed on 22 November 2021). [CrossRef]

35. Nicol, T.; Lefeuvre, C.; Serri, O.; Pivert, A.; Joubaud, F.; Dubee, V.; Kouatchet, A.; Ducancelle, A.; Lunel-Fabiani, F.; Le Guillou-Guillemette, H. Assessment of SARS-CoV-2 serological tests for the diagnosis of COVID-19 through the evaluation of three immunoassays: Two automated immunoassays (Euroimmun and Abbott) and one rapid lateral flow immunoassay (NG Biotech). *J. Clin. Virol.* **2020**, *129*, 104511. [CrossRef]

36. Schroeder, H.W., Jr.; Cavacini, L. Structure and function of immunoglobulins. *J. Allergy Clin. Immunol.* **2010**, *125*, S41–S52. [CrossRef] [PubMed]

37. Kaewpoowat, Q.; Salazar, L.; Aguilera, E.; Wootton, S.H.; Hasbun, R. Herpes simplex and varicella zoster CNS infections: Clinical presentations, treatments and outcomes. *Infection* **2016**, *44*, 337–345. [CrossRef]

38. Bunyaratavej, S.; Prasertyothin, S.; Leeyaphan, C. Atypical Presentation of Recurrent Varicella Zoster Virus Infection: A Case Report and Review of the Literature. *Southeast Asian J. Trop. Med. Public Health* **2015**, *46*, 27–29. [PubMed]

39. Freer, G.; Pistello, M. Varicella-zoster virus infection: Natural history, clinical manifestations, immunity and current and future vaccination strategies. *New Microbiol.* **2018**, *41*, 95–105.

40. van Loon, A.M.; van der Logt, J.T.; Heessen, F.W.; Heeren, M.C.; Zoll, J. Antibody-capture enzyme-linked immunosorbent assays that use enzyme-labelled antigen for detection of virus-specific immunoglobulin M, A and G in patients with varicella or herpes zoster. *Epidemiol. Infect.* **1992**, *108*, 165–174. [CrossRef]
41. Zerboni, L.; Sen, N.; Oliver, S.L.; Arvin, A.M. Molecular mechanisms of varicella zoster virus pathogenesis. *Nat. Rev. Microbiol.* **2014**, *12*, 197–210. [CrossRef] [PubMed]
42. Oliver, S.L.; Yang, E.; Arvin, A.M. Varicella-Zoster Virus Glycoproteins: Entry, Replication, and Pathogenesis. *Curr. Clin. Microbiol. Rep.* **2016**, *3*, 204–215. [CrossRef]
43. Grahn, A.; Studahl, M.; Nilsson, S.; Thomsson, E.; Backstrom, M.; Bergstrom, T. Varicella-zoster virus (VZV) glycoprotein E is a serological antigen for detection of intrathecal antibodies to VZV in central nervous system infections, without cross-reaction to herpes simplex virus 1. *Clin. Vaccine Immunol.* **2011**, *18*, 1336–1342. [CrossRef] [PubMed]
44. Min, S.W.; Kim, Y.S.; Nahm, F.S.; Yoo, D.H.; Choi, E.; Lee, P.B.; Choo, H.; Park, Z.Y.; Yang, C.S. The positive duration of varicella zoster immunoglobulin M antibody test in herpes zoster. *Medicine* **2016**, *95*, e4616. [CrossRef] [PubMed]
45. Enders, G. Serodiagnosis of Varicella-Zoster virus infection in pregnancy and standardization of the ELISA IgG and IgM antibody tests. *Dev. Biol. Stand.* **1982**, *52*, 221–236. [PubMed]
46. Iltis, J.P.; Achilli, G.; Madden, D.L.; Sever, J.L. Serologic study by enzyme-linked immunosorbent assay of the IgM antibody response in the patas monkey following experimental simian varicella virus infection. *Diagn. Immunol.* **1984**, *2*, 137–142.
47. Dobec, M.; Bossart, W.; Kaeppeli, F.; Mueller-Schoop, J. Serology and serum DNA detection in shingles. *Swiss Med. Wkly.* **2008**, *138*, 47–51.
48. Medic, S.; Petrovic, V.; Milosevic, V.; Lozanov-Crvenkovic, Z.; Brkic, S.; Andrews, N.; de Ory, F.; Anastassopoulou, C. Seroepidemiology of varicella zoster virus infection in Vojvodina, Serbia. *Epidemiol. Infect.* **2018**, *146*, 1593–1601. [CrossRef]
49. Hellstern, P.; Solheim, B.G. The Use of Solvent/Detergent Treatment in Pathogen Reduction of Plasma. *Transfus. Med. Hemoth.* **2011**, *38*, 65–70. [CrossRef] [PubMed]

pathogens

Article

Antibodies to Highly Pathogenic A/H5Nx (Clade 2.3.4.4) Influenza Viruses in the Sera of Vietnamese Residents

Tatyana Ilyicheva [1,2,*], Vasily Marchenko [1], Olga Pyankova [1], Anastasia Moiseeva [1], Tran Thi Nhai [3], Bui Thi Lan Anh [3], Trinh Khac Sau [3], Andrey Kuznetsov [3], Alexander Ryzhikov [1] and Rinat Maksyutov [1]

[1] State Research Center of Virology and Biotechnology "Vector", Rospotrebnadzor, Koltsovo, Novosibirsk 630559, Russia; marchenko_vyu@vector.nsc.ru (V.M.); pyankova_og@vector.nsc.ru (O.P.); chalaya_aa@vector.nsc.ru (A.M.); ryzhik@vector.nsc.ru (A.R.); maksyutov_ra@vector.nsc.ru (R.M.)
[2] Department Natural Sciences, Novosibirsk State University, Novosibirsk 630090, Russia
[3] Russian-Vietnamese Tropical Research and Technology Centre, Hanoi 650000, Vietnam; tbnnhai@yahoo.com (T.T.N.); lananhrus@yahoo.com (B.T.L.A.); sau_tk@yahoo.com (T.K.S.); tropcenterhanoi@mail.ru (A.K.)
* Correspondence: ilyichev@mail.ru

Abstract: To cause a pandemic, an influenza virus has to overcome two main barriers. First, the virus has to be antigenically new to humans. Second, the virus has to be directly transmitted from humans to humans. Thus, if the avian influenza virus is able to pass the second barrier, it could cause a pandemic, since there is no immunity to avian influenza in the human population. To determine whether the adaptation process is ongoing, analyses of human sera could be conducted in populations inhabiting regions where pandemic virus variant emergence is highly possible. This study aimed to analyze the sera of Vietnamese residents using hemagglutinin inhibition reaction (HI) and microneutralization (MN) with A/H5Nx (clade 2.3.4.4) influenza viruses isolated in Vietnam and the Russian Federation in 2017–2018. In this study, we used sera from 295 residents of the Socialist Republic of Vietnam collected from three groups: 52 samples were collected from households in Nam Dinh province, where poultry deaths have been reported (2017); 96 (2017) and 147 (2018) samples were collected from patients with somatic but not infectious diseases in Hanoi. In all, 65 serum samples were positive for HI, at least to one H5 virus used in the study. In MN, 47 serum samples neutralizing one or two viruses at dilutions of 1/40 or higher were identified. We postulate that the rapidly evolving A/H5Nx (clade 2.3.4.4) influenza virus is possibly gradually adapting to the human host, insofar as healthy individuals have antibodies to a wide spectrum of variants of that subtype.

Keywords: highly pathogenic avian influenza virus; H5N6 (clade 2.3.4.4); human sera

Citation: Ilyicheva, T.; Marchenko, V.; Pyankova, O.; Moiseeva, A.; Nhai, T.T.; Lan Anh, B.T.; Sau, T.K.; Kuznetsov, A.; Ryzhikov, A.; Maksyutov, R. Antibodies to Highly Pathogenic A/H5Nx (Clade 2.3.4.4) Influenza Viruses in the Sera of Vietnamese Residents. *Pathogens* **2021**, *10*, 394. https://doi.org/10.3390/pathogens10040394

Academic Editors: Philipp A. Ilinykh, Kai Huang and Xuguang Li

Received: 9 February 2021
Accepted: 23 March 2021
Published: 25 March 2021

Publisher's Note: MDPI stays neutral with regard to jurisdictional claims in published maps and institutional affiliations.

1. Introduction

Humanity has acquired knowledge to control most anthroponotic infections, such as measles, poliomyelitis, smallpox, and mumps. However, recently, problems related to zoonotic human infections have emerged. Since the natural reservoirs of these pathogens are unlimited, they are difficult or often impossible to control.

Influenza A virus belongs to the genus *Alphainfluenzaviruses* of the *Orthomyxoviridae* family and has a segmented genome consisting of single-stranded RNA segments of negative polarity [1]. Influenza A viruses are divided into subtypes (serotypes) based on the genetic and antigenic characteristics of its two surface glycoproteins, hemagglutinin (HA) and neuraminidase (NA). The nomenclature of these viruses is based on a combination of the HA (H1–H18) and NA (N1–N11) subtypes. Wild waterfowl are a natural reservoir for all influenza A subtypes [2], except for H17N10 and H18N11, which were recently found in bats [3,4]. Influenza A viruses can also be detected in a wide variety of hosts including humans, swine, horses, dogs, cats, and sea mammals.

Pandemic influenza A virus appears in the human population every 10–30 years. There is currently no immunity to these viruses; therefore, the resultant pandemics cause high morbidity and often high mortality. It is supposed that pandemic influenza A viruses emerge because of the reassortment of viruses in humans and animals or the adaptation of zoonotic viruses to humans [5].

Reassortment occurs when two strains of influenza A virus coinfect one cell and can cause procreation of the new reassortant virus, which contains a new set of genes [6]. The influenza pandemics in 1957 and 1968 were caused by reassortant viruses containing genes of influenza A viruses of humans and birds [7]. The first pandemic of the 21st century was caused by a virus that emerged after multiple reassortments of human-, avian-, and swine-origin influenza A viruses [8].

The views on the cause of the 1918 pandemic (the so-called "Spanish flu") differ among experts. Some support the idea that the virus was directly introduced into the human population (without reassortment), while others believe that the pandemic virus emerged after multiple genome reassortments of avian and mammalian, and possibly swine and/or human, viruses that had emerged during the years preceding the pandemic of 1918 [9].

Regardless of the exact mechanism of the emergence of a new virus variant, there may be a certain period before the pandemic begins that is needed by the virus for optimal adaptation to the human host [10].

If this assumption is correct, then an understanding of whether the adaptation process is ongoing is possible by analyzing antibody levels in the sera of human populations inhabiting regions where the emergence of pandemic virus variants is most likely.

The objective of this research was to analyze the sera of Vietnamese residents using the hemagglutinin inhibition (HI) test and virus microneutralization (MN) with avian influenza viruses isolated in Vietnam and the Russian Federation in 2017–2018.

2. Results

For sera analyses, we selected influenza A virus strains that were isolated from poultry in Vietnam in 2018 and poultry and wild birds in Russia in 2017–2018. These strains were selected based on a phylogenetic analysis, which revealed a high degree of identity between the strains isolated in Russia and Vietnam (Figure 1). It was determined that the A/chicken/Nghe An/27VTC/2018 (H5N6), A/chicken/Nghe An/01VTC/2018 (H5N6), and A/common gull/Saratov/1676/2018 (H5N6) strains belonged to the genetic clade 2.3.4.4h. At that time, the strain A/chicken/Kostroma/1718/2017 (H5N2) was selected for comparison as a representative of clade 2.3.4.4b, which circulated widely in Eurasia. The study of antigenic properties is consistent with the phylogenetic analysis. Data are presented in Table 1.

As shown in Table 1, the A/chicken/NgheAn/27VTC/2018 (H5N6) and A/chicken/NgheAn/01VTC/2018 (H5N6) viruses cross-reacted with sera against A/common gull/Saratov/1676/2018 (H5N6) as effectively as the homological virus (the reciprocal of the titer in the HI test is 320). Vietnamese strains did not react with other ferret reference sera obtained against the A/H5N6 viruses isolated in Russia.

For the analysis of sera against influenza virus subtype A/H5, we selected the following viruses: low pathogenic A/chicken/Kostroma/1718/2017 (H5N2) virus (LPAI) and highly pathogenic A/common gull/Saratov/1676/2018, A/chicken /Nghe An/01VTC/2018, and A/chicken/Nghe An/27VTC/2018 (H5N6) viruses (HPAI).

The sera were tested using the HI test with human vaccine viruses, highly and lowly pathogenic viruses of the A/H5 subtype, highly pathogenic A/H7N9 virus, and lowly pathogenic A/H9N2 virus.

Testing with vaccine strains showed that antibodies with significant titers (40 and higher) to the A/Michigan/45/2015 (H1N1pdm09) virus were detected in 29 samples (14, 2, and 13 from three groups, respectively). Antibodies to the A/Singapore/INFIMH-16-0019/2016 (H3N2) vaccine virus were detected in 55 samples (14, 6, and 35 from each group, respectively) (data not shown).

None of the tested serum samples reacted in HI or MN with A/Anhui/1/2013 (H7N9) virus at dilutions of 1:40 or higher. Four samples had a titer of 20 in the HI test, but not in MN. Fifty-seven samples reacted with the A/chicken/Primorsky Krai/03/2018 (H9N2) virus in the HI test, of which 0, 1, and 56 samples were from the three groups, respectively (data not shown).

The results of the sera analyses in HI and MN for A/H5Nx (2.3.4.4.) avian influenza viruses are presented in Table 2. Data are presented only for samples that were positive in at least one study.

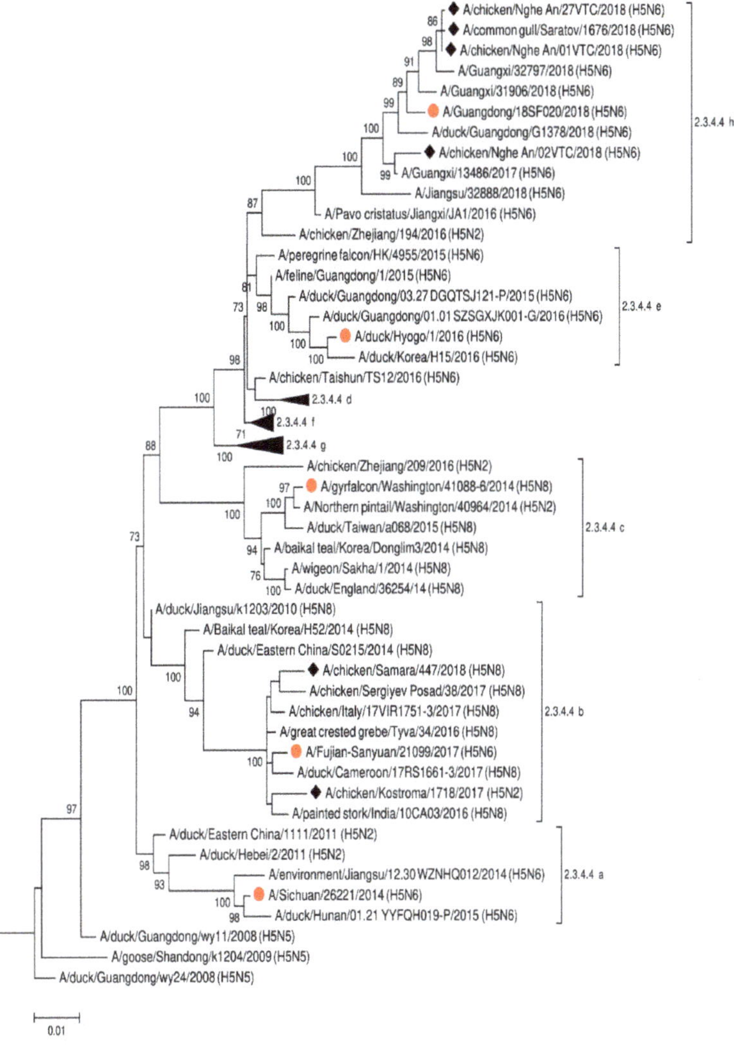

Figure 1. The phylogenetic tree for HA of A(H5Nx) influenza viruses. Viruses used in this study are indicated by rhombs. Candidate vaccine viruses are indicated by circles (according to WHO recommendations https://www.who.int/influenza/vaccines/virus/characteristics_virus_vaccines/en/, accessed on 5 March 2021).

Table 1. HI test of H5Nx clade 2.3.4.4 viruses with ferret reference antisera and horse red blood cells.

Virus	Subtype	Clade	Reverse Titer with Antisera									
			A/duck/ England/ 36254/2014	A/Northern Pintail/WA/ 40964/2014	A/Sichuan/ 26221/2014 RG42A	A/gyrfalcon/ WA/41088/ 2014 RG43A	A/great crested grebe/Tyva/ 34/2016	A/wigeon/ Sakha/1/2014	A/chicken/ SergiyevPosad/ 38/2017	A/chicken/ Kostroma/ 1718/2017	A/common gull/Saratov/ 1676/2018	A/chicken/ Nghe An/ 27VTC/2018
A/duck/ England/ 36254/2014	H5N8	2.3.4.4c	**640**	5120	5120	640	1280	1280	1280	2560	NA	NA
A/Northern Pintail/WA/ 40964/2014	H5N2	2.3.4.4c	320	**5120**	2560	2560	640	640	160	1280	<20	<20
A/Sichuan/ 26221/2014 RG42A	H5N6	2.3.4.4a	640	5120	**5120**	320	640	320	320	1280	<20	NA
A/gyrfalcon/ WA/41088/ 2014 RG43A	H5N8	2.3.4.4c	640	5120	1280	**5120**	1280	1280	320	1280	<20	<20
A/great crested grebe/Tyva/ 34/2016	H5N8	2.3.4.4b	320	5120	2560	2560	**1280**	640	640	1280	<20	<20
A/wigeon/ Sakha/1/2014	H5N8	2.3.4.4c	640	10,240	2560	5120	640	**640**	640	2560	<20	<20
A/chicken/ SergiyevPosad/ 38/2017	H5N8	2.3.4.4b	160	2560	2560	2560	320	640	**160**	1280	<20	<20
A/chicken/ Kostroma/ 1718/2017	H5N2	2.3.4.4b	320	10,240	2560	2560	640	640	320	**5120**	<20	<20
A/common gull/Saratov/ 1676/2018	H5N6	2.3.4.4h	<20	<20	80	<20	<20	<20	<20	<20	**320**	80
A/chicken/ NgheAn/ 27VTC/2018	H5N6	2.3.4.4h	<20	320	40	20	<20	<20	<20	<20	320	**80**
A/chicken/ NgheAn/ 01VTC/2018	H5N6	2.3.4.4h	NA	640	80	<20	<20	<20	<20	<20	320	160
A/chicken/ Vietnam/ NCVD-15A59/2015	H5N6	2.3.4.4f	320	10,240	2560	2560	640	640	160	640	40	<20

HI titers of sera with a homologous viruses are highlighted in bold.

Table 2. MN and HI test of human sera with H5 avian influenza viruses.

Human Serum Sample	Group	A/chicken/ NgheAn/01VTC/2018 (H5N6) 2.3.4.4		A/chicken/ Kostroma1718/2017 (H5N2) 2.3.4.4		A/chicken/ NgheAn/27VTC/2018 (H5N6) 2.3.4.4		A/common Gull/Saratov/1676/2018 (H5N6) 2.3.4.4	
		HI	MN	HI	MN	HI	MN	HI	MN
20	1			≥160					
56	2			40					
60	2	40							
82	2			40					
100	2	40		40					
105	2	80							
128	2	40		40					
134	2			40					
200	3	≥160		≥160		≥160	160	160	80
203	3	80	80	160		160	80	≥160	80
209	3	≥160	80	160		≥160	160	≥160	160
212	3	≥160		160		80	80	≥160	160
213	3	≥160		160		160	80	≥160	80
214	3	≥160		160		160	80	≥160	80
221	3	80		≥160		80	80	≥160	160
222	3	40		160		40	80	80	80
231	3			40					
232	3			40					
235	3	80		160		160	320	≥160	80
237	3	80		≥160		160	320	≥160	160
239	3	80		160		40	80	160	160
240	3	≥160		≥160		160	80	≥160	320
241	3	≥160		160		160	80	≥160	320
250	3	80		≥160		80	80	160	160

Table 2. *Cont.*

Human Serum Sample	Group	A/chicken/ NgheAn/01VTC/2018 (H5N6) 2.3.4.4		A/chicken/ Kostroma1718/2017 (H5N2) 2.3.4.4		A/chicken/ NgheAn/27VTC/2018 (H5N6) 2.3.4.4		A/common Gull/Saratov/1676/2018 (H5N6) 2.3.4.4	
		HI	MN	HI	MN	HI	MN	HI	MN
251	3	40		160					
255	3	≥160		≥160		≥160	80	≥160	320
257	3	≥160		≥160		80	80	80	160
258	3	≥160		≥160		160	80	≥160	160
259	3	≥160		≥160		160	160	≥160	160
268	3	≥160		≥160		80	160	≥160	80
270	3	80		≥160		40	80	80	
271	3	≥160		≥160		80	160	≥160	80
272	3	≥160		≥160		80	160	≥160	
273	3	80		≥160		80	80	≥160	320
274	3	80		≥160		80	80	≥160	80
275	3	≥160		≥160		≥160	160	≥160	
277	3	40		40					
278	3	≥160		160		80	160	≥160	80
281	3	≥160		≥160		80	160	160	
283	3	≥160		80		80	80	≥160	
284	3	≥160		≥160		≥160	160	≥160	
285	3			≥160		≥160	320	≥160	
286	3	≥160		≥160		80	160	≥160	
287	3	≥160		80		80	80	≥160	
288	3			80		40		≥160	
289	3	≥160		≥160		160	160	≥160	
290	3	80		80		80	80	80	
291	3	80		160		160	80	≥160	

Table 2. *Cont.*

Human Serum Sample	Group	A/chicken/ NgheAn/01VTC/2018 (H5N6) 2.3.4.4		A/chicken/ Kostroma1718/2017 (H5N2) 2.3.4.4		A/chicken/ NgheAn/27VTC/2018 (H5N6) 2.3.4.4		A/common Gull/Saratov/1676/2018 (H5N6) 2.3.4.4	
		HI	MN	HI	MN	HI	MN	HI	MN
292	3	80		$\geq$160		160	80	$\geq$160	
293	3	80		160		80	80	$\geq$160	
294	3	80		$\geq$160		80	80	$\geq$160	
295	3	80		160		80	80	$\geq$160	
296	3	80		160		40		40	
297	3	40		40				40	
298	3	40		80					
299	3	80		80		80	160	80	
300	3	80		80		80	160	80	
303	3	40		80		40		80	
304	3	40		80				160	
307	3	40		80				40	160
308	3	80		80		80	80	80	160
309	3	40		$\geq$160		40	80	40	
310	3	40		80		40		40	
334	3	80		80	80	80	160	40	
335	3	80		80		80	160	80	80
338	3	80		80		80	160	40	80
339	3	80		80		80	160	80	
Total number of positive sera		59	2	65	1	51	47	54	27

Reciprocal serum titers are shown for the selected sera positive in HI and MN simultaneously at least for one virus. The positive serum in MN $\geq$ 40, the positive serum in HI test $\geq$ 40. Each serum was tested three times in HI test with horse red blood cells; the results differed by no more than 2 times, and a lower value was taken for the serum titer. Sera from recovered ferrets infected with analyzed virus strain were used as a positive control. Negative control was represented by sera of non-immune ferrets and human sera without antibodies to avian influenza virus. Negative control titer in all tests > 20, positive control titer from 320 to 640.

As shown in Table 2, only one serum from the first group reacted in the HI test with one of the viruses, A/chicken/Kostroma/1718/2017 (H5N2) 2.3.4.4. Seven serum samples from the second group reacted in the HI test with one or two A/H5 viruses. The largest number of positive serum samples was in the third group: 55 samples reacted in the HI test and two samples in MN with the A/chicken/NgheAn/01VTC/2018 (H5N6) virus, 59 serum samples were positive in the HI test and only one in MN with A/chicken/Kostroma/1718/2017 (H5N2) virus, 51 serum samples in HI and 47 in MN with A/chicken/NgheAn/27VTC/2018 (H5N6) virus, and 54 serum samples were positive in HI and 27 in MN with the A/common gull/Saratov/1676/2018 (H5N6) virus.

3. Discussion

The first documented outbreak of human infection with the avian influenza A/H5 virus occurred in Hong Kong in 1997. Since then, A(H5N1) has caused diseases in 861 people, and 455 cases were fatal (data up to 20 September 2020). In Vietnam, 127 confirmed cases of human infection with the A(H5N1) virus were detected between 2003 and 2014, 64 of which were fatal. No human cases of highly pathogenic avian influenza A/H5 have been reported in Vietnam since 2015 [11]. However, outbreaks in poultry and wild birds have been reported in Vietnam to date, including those caused by the most rapidly evolving H5N6 subtype [12].

Today, HPAI H5N6 is one of the few subtypes of the avian influenza virus that can infect humans [13]. Studies have shown that the H5N6 virus originated from a common precursor strain of the clade 2.3.4.4 subtype H5 as a result of reassortment with the A/duck/Guangxi/2281/2007 (H6N6) strain [13–16]. However, there is evidence that the origin of the H5N6 viruses followed other evolutionary paths [17]. According to one of the hypotheses, this virus appeared in the period from 2010 to 2012 as a result of reassortment of the H5N2 virus of clade 2.3.4.4 with the A/duck/Guangxi/2281/2007 (H6N6) strain, followed by reassortment of the six internal genes with the H5N1 influenza virus of clade 2.3.2.1c isolated from chickens [14]. The H5N6 virus, the so-called "reassortant A", which developed along this pathway, circulates in Xinjiang, Jilin, and Northern China [14,15,18]. In 2013, the virus spread to Western China, where it caused outbreaks in Sichuan, Vietnam, and Laos [14,19].

Another variant of the H5N6 influenza virus, named "reassortant B," appeared in 2013 because of reassortment of H6N6 viruses with H5N8 viruses in clade 2.3.4.4 and subsequent resorting of genes with H5N1 viruses in clade 2.3.2.1c [14]. This virus also circulated in China, Vietnam, and Laos. Two years later, reassortant B underwent repeated reassortment with an influenza virus of the H9N2 subtype, resulting in a new variant of the H5N6 virus, reassortant C. The circulation of this variant was reported in the Yunnan and Guangdong provinces of China [17,18]. Regardless of the evolutionary way in which these reassortants emerged, it is known that all of them caused infectious diseases among humans [14–16]. Since 2014, H5N6 viruses (2.3.4.4) have caused 27 human infections, resulting in 15 deaths in China [20]. Most human cases are associated with A or B reassortants.

In 2014, H5N6 spread across Laos and Vietnam, resulting in huge economic losses due to outbreaks in poultry [18,21]. In 2016, H5N6 caused several outbreaks in Japan, Myanmar, and the Republic of Korea [14,15,21]. In 2017, outbreaks of H5N6 were reported in Taiwan and the Philippines. In addition, with wild migratory birds, HPAI H5N6 was introduced into Europe during this time. Outbreaks were reported in Greece, Germany, the Netherlands, and Switzerland [14–17,22]. To date, H5N6 circulation is limited to Asia and Europe.

In addition to influenza viruses of the A/H5 subtype, the H6, H7, H9, and H10 viruses also have pandemic potential, since there is confirmed evidence that they can cross the interspecies barrier and infect humans [23]. For example, highly pathogenic A(H7N9) viruses caused 1568 confirmed human cases, 616 of which were fatal (data up to 20 September 2020) [24].

To cause a pandemic, the virus has to overcome two main obstacles. First, the pathogen must be antigenically new to humans to ensure that herd immunity does not impede the rapid spread of the virus. Second, the virus must be effectively transmitted from person to person [25]. Thus, if the avian influenza virus is able to pass the second barrier, it could cause a pandemic, since there is no immunity to avian influenza virus in the human population.

Vietnam is considered one of the hotspots for the emergence of influenza viruses with epidemic and pandemic potential [26]. A/H5 highly pathogenic avian influenza viruses have been endemic in Vietnamese poultry for over a decade and a half. It has been shown that in the northern provinces of Vietnam, more than 30% of poultry have antibodies to the influenza A/H5 virus [26]. The Global Influenza Surveillance and Response System (GISRS) tracks the emergence of viruses with pandemic potential. Within the framework of GISRS, this work is being carried out at the WHO Reference Laboratory for H5 on the basis of the State Research Center of VB "Vector" of Rospotrebnadzor [27] and in the Socialist Republic of Vietnam [28–30].

However, during the adaptation process, virus variants may not cause clinical manifestations and may remain unnoticed during monitoring activities. It is even more difficult to detect the adaptation of lowly pathogenic avian influenza viruses to humans. In such cases, adaptation can be assessed by analyzing sera collected from residents of areas with a high risk of avian influenza.

In the present study, we investigated the sera of Vietnamese residents using MN and HI with highly pathogenic, lowly pathogenic, and vaccine influenza A viruses H5N2, H5N6, H7N9, H9N2, A/H1N1pdm09, and A/H3N2 subtypes. The analyses showed that only 10% of sera had significant titers in the HI test with the A/H1N1pdm09 vaccine virus and close to 20% with the A/H3N2 vaccine virus. This indicates a low level of population immunity against seasonal influenza, which represents a serious risk, since the probability of human infection with a seasonal and highly pathogenic influenza virus is increasing, and this event could lead to the emergence of a reassortant virus with new antigenic features and the ability to be transmitted from person to person.

In the tested sera, we did not find significant titers of antibodies against the A/Anhui/1/2013 (H7N9) virus; only four serum samples out of 295 had a titer of 1:20 in HI, but not in MN. With the A/chicken/PrimorskyKrai/03/2018 (H9N2) virus, 57 serum samples were positive. This is consistent with previous data on the wide circulation of the A(H9N2) virus in Vietnam, especially in live poultry markets [31], and relatively high antibody levels to this virus serotype in human sera [28]. Current concerns about H9N2 viruses are related to their ability to reassort with other avian influenza viruses, resulting in highly and lowly pathogenic viruses that can cross species barriers and infect humans [32].

Among all investigated viruses of the A/H5 subtype, only two viruses, A/chicken/Nghe An/27VTC/2018 (H5N6) 2.3.4.4 and A/common gull/Saratov/1676/2018 (H5N6) 2.3.4.4, had a significant number of positive serum samples in HI and MN. For other viruses, we found mostly anti-hemagglutination antibodies, but not neutralizing virus in cell cultures. However, the absence of neutralizing antibodies does not mean that humans have not been infected. Thus, Li et al. showed that after mice were infected with wildtype A/H7N9 virus, all animals after 14 days had high titers of HI antibodies in their sera, but not virus-neutralizing antibodies. At the same time, the recombinant virus, which contained genes for internal proteins from the PR8 strain, and the HA and NA genes from A/H7N9, induced significantly higher antibody levels in sera, detected in both HI and MN. The authors concluded that internal proteins of the A/H7N9 virus can influence the humoral immune response of the host [33].

We analyzed all sera in the HI test with horse, goose, and turkey red blood cells and demonstrated the highest titers with horse red blood cells, in accordance with [34]. It is not clear why, in the present study, many serum samples from the third group reacted with different A/H5 viruses in HI (but not in MN). Thus, 45 serum samples were positive in the HI test and all were included in the study of A/H5 viruses. At the same time, these viruses in most cases did not react in HI with heterologous ferret reference sera. However, antibodies may appear because of human infection with various A/H5 pathogens. However, in our opinion, it is highly probable that the presence of antibodies to different A/H5 viruses in the same serum samples can be explained by the fact that A/H5 viruses induce in humans (but not in ferrets) a wide range of anti-H5 cross-reactive antibodies. This may be because of frequent human contact with various virus subtypes, while, to obtain a reference serum, animals that do not have antibodies to any influenza virus are selected.

To understand this phenomenon and, more importantly, the process of avian influenza virus adaptation to humans, it is necessary to continue studies of the circulation of viruses in domestic and wild birds, as well as the sera of people living in "hotspot" regions of pandemic potential virus emergence. From the results of the present research, we can postulate that the rapidly evolving A/H5Nx influenza viruses (clade 2.3.4.4) are gradually adapting to human hosts, in so far as healthy individuals have neutralizing and anti-hemagglutinating antibodies to a wide range of viruses of this subtype. The number and profiles of serum samples tested in this work were limited, so our results may be misleading; thus, we cannot say with confidence that the process of adaptation of H5 avian viruses to humans is underway. Nevertheless, our results allow us to make such an assumption.

It should be noted that in the human population, there are diseases that are caused by pathogens that previously circulated only among animals: SARS, the Middle East respiratory syndrome, coronavirus disease 2019 (COVID-19), acquired immune deficiency syndrome, and pandemics caused by Chikungunya and Zika viruses [35]. It is believed that, in the future, new viruses that are pathogenic to humans will emerge, owing to the presence of an unlimited natural reservoir of animal viruses. One of the most dangerous among them includes highly pathogenic influenza viruses. It is possible that pandemics similar to the Spanish flu will emerge [36]. Therefore, it is important to conduct comprehensive surveillance of avian and mammalian influenza viruses, including monitoring human sera for the presence of antibodies to animal influenza viruses.

4. Materials and Methods

Sera. The blood serum research was approved by the Ethics Committee IRB 00001360, affiliated with SRC VB Vector (No.2 d.d. Protocol, May 2008). The study used blood sera collected from residents of the Socialist Republic of Vietnam; 52 samples (No 1-52) were collected from private households in Nam Dinh province, where poultry deaths have been reported (2017) [37]; 96 samples (No 53-148) were collected in Hanoi from healthy donors (2017), and 147 samples (No 200-346) were collected from patients with non-communicable diseases in hospitals in Hanoi (2018). Blood samples were collected, on condition of anonymity, from individuals of different age groups: 18–55 years (90%), 56–64 years (8%) and 65 years and older (2%).

Before HI testing, sera were treated with receptor-destroying enzyme and hemadsorbed on horse RBCs, before MN sera were heat-inactivated for 30 min at 56 °C, as described by [34,38].

Viruses. A/Michigan/45/2015 (H1N1) pdm09 and A/Singapore/INFIMH-16-0019/2016 (H3N2) vaccine influenza viruses were kindly provided by the WHO Collaborating Center in Atlanta, United States. The WHO Collaborating Center in Beijing, China, kindly provided the A/Anhui/01/2013 (H7N9) virus. Virus A/chicken/Kostroma/1718/2017 (H5N2) [39], virus A/common gull/Saratov/1676/2018 (H5N6) [40,41], A/chicken/PrimorskyKrai/03/2018 (H9N2), A/chicken/NgheAn/01VTC/2018 (H5N6), and A/chicken/NgheAn/27VTC/2018 (H5N6) were isolated by the authors. A maximum-likelihood tree based on the Hasegawa–Kishino–Yano model was built using MEGA 6.06 software (http://www.megasoftware.net/, accessed on 5 March 2021) with 1000 bootstrap replicates.

HI test and MN. The hemagglutination inhibition (HI) test and microneutralization (MN) method were performed as described by [38,41]. Horse, goose, and turkey red blood cells were used in HI tests. Highest titers were obtained with horse red blood cells. Each serum was tested three times in the HI test with horse red blood cells, and the results differed by no more than 2 times. A lower value was taken for the serum titer. Sera from recovered ferrets infected with analyzed virus strain were used as a positive control. Negative control was represented by sera of non-immune ferrets and human sera without antibodies to avian influenza virus.

Funding

The study was conducted at the expense of targeted subsidies by order of the Government of the Russian Federation of July 13, 2019 No. 1536-r. and by State Assignment no. 1/21 and no. 3/21 (SRC VB "Vector").

Author Contributions: Conceptualization, T.I., A.K. and A.R.; Formal analysis, V.M. and T.K.S.; Investigation, O.P., A.M., T.T.N. and B.T.L.A.; Methodology, V.M. and T.K.S.; Project administration, A.K. and A.R.; Supervision, A.K. and R.M.; Writing—original draft, T.I.; Writing—review and editing, T.I. and A.R. All authors have read and agreed to the published version of the manuscript.

Institutional Review Board Statement: The blood serum research was approved by the Ethics Committee IRB 00001360, affiliated with SRC VB Vector (No.2 d.d. Protocol, May 2008).

Informed Consent Statement: Informed consent was obtained from all subjects involved in the study.

Data Availability Statement: Data sharing is not applicable to this article.

Conflicts of Interest: The authors declare no conflict of interest.

References

1. Krammer, F.; Smith, G.J.D.; Fouchier, R.A.M.; Peiris, M.; Kedzierska, K.; Doherty, P.C.; Palese, P.; Shaw, M.L.; Treanor, J.; Webster, R.G.; et al. Influenza. *Nat. Rev. Dis. Primers* **2018**, *4*, 1–21. [CrossRef] [PubMed]
2. Russell, C.J.; Hu, M.; Okda, F.A. Influenza Hemagglutinin Protein Stability, Activation, and Pandemic Risk. *Trends Microbiol.* **2018**, *26*, 841–853. [CrossRef]
3. Tong, S.; Li, Y.; Rivailler, P.; Conrardy, C.; Castillo, D.A.; Chen, L.M.; Recuenco, S.; Ellison, J.A.; Davis, C.T.; York, I.A.; et al. A distinct lineage of influenza A virus from bats. *Proc. Natl. Acad. Sci. USA* **2012**, *109*, 4269–4274. [CrossRef] [PubMed]
4. Tong, S.; Zhu, X.; Li, Y.; Shi, M.; Zhang, J.; Bourgeois, M.; Yang, H.; Chen, X.; Recuenco, S.; Gomez, J.; et al. New world bats harbor diverse influenza A viruses. *PLoS Pathog.* **2013**, *9*, e1003657. [CrossRef]
5. Hay, A.J.; Gregory, V.; Douglas, A.R.; Yi, P.L. The evolution of human influenza viruses. *Philos. Trans. R. Soc. B Biol. Sci.* **2001**, *356*, 1861–1870. [CrossRef]
6. Short, K.R.; Kedzierska, K.; van de Sandt, C.E. Back to the Future: Lessons Learned From the 1918 Influenza Pandemic. *Front. Cell. Infect. Microbiol.* **2018**, *8*, 343. [CrossRef]
7. Scholtissek, C.; Rohde, W.; Von Hoyningen, V.; Rott, R. On the origin of the human influenza virus subtypes H2N2 and H3N2. *Virology* **1978**, *87*, 13–20. [CrossRef]
8. Smith, G.J.D.; Vijaykrishna, D.; Bahl, J.; Lycett, S.J.; Worobey, M.; Pybus, O.G.; Ma, S.K.; Cheung, C.L.; Raghwani, J.; Bhatt, S.; et al. Origins and evolutionary genomics of the 2009 swineorigin H1N1 influenza A epidemic. *Nature* **2009**, *459*, 1122–1125. [CrossRef]
9. Worobey, M.; Han, G.Z.; Rambaut, A. Genesis and pathogenesis of the 1918 pandemic H1N1 influenza A virus. *Proc. Natl. Acad. Sci. USA* **2014**, *111*, 8107–8112. [CrossRef]
10. Morens, D.M.; Taubenberger, J.K. Influenza Cataclysm, 1918. *N. Engl. J. Med.* **2018**, *379*, 2285–2287. [CrossRef]
11. WHO H5N1. Cumulative Number of Confirmed Human Cases of Avian Influenza A(H5N1) Reported to WHO. Available online: http://www.who.int/influenza/human_animal_interface/H5N1_cumulative_table_archives/en/ (accessed on 11 March 2021).
12. Tsunekuni, R.; Sudo, K.; Nguyen, P.T.; Luu, B.D.; Phuong, T.D.; Tan, T.M.; Nguyen, T.; Mine, J.; Nakayama, M.; Tanikawa, T.; et al. Isolation of highly pathogenic H5N6 avian influenza virus in Southern Vietnam with genetic similarity to those infecting humans in China. *Transbound. Emerg. Dis.* **2019**, *66*, 2209–2217. [CrossRef] [PubMed]
13. Thanh, H.D.; Tran, V.T.; Nguyen, D.T.; Hung, V.K.; Kim, W. Novel reassortant H5N6 highly pathogenic influenza A viruses in Vietnamese quail outbreaks. *Comp. Immunol. Microbiol. Infect. Dis.* **2018**, *56*, 45–57. [CrossRef]
14. Bi, Y.; Chen, Q.; Wang, Q.; Chen, J.; Jin, T.; Wong, G.; Quan, C.; Liu, J.; Wu, J.; Yin, R.; et al. Genesis, Evolution and Prevalence of H5N6 Avian Influenza Viruses in China. *Cell Host Microbe* **2016**, *20*, 810–821. [CrossRef] [PubMed]
15. Bi, Y.; Liu, H.; Xiong, C.; Di, L.; Shi, W.; Li, M.; Liu, S.; Chen, J.; Chen, G.; Li, Y.; et al. Novel avian influenza A (H5N6) viruses isolated in migratory waterfowl before the first human case reported in China, 2014. *Sci. Rep.* **2016**, *6*, 29888. [CrossRef]
16. Takemae, N.; Tsunekuni, R.; Sharshov, K.; Tanikawa, T.; Uchida, Y.; Ito, H.; Soda, K.; Usui, T.; Sobolev, I.; Shestopalov, A.; et al. Five distinct reassortants of H5N6 highly pathogenic avian influenza A viruses affected Japan during the winter of 2016–2017. *Virology* **2017**, *512*, 8–20. [CrossRef] [PubMed]

17. Yang, L.; Zhu, W.; Li, X.; Bo, H.; Zhang, Y.; Zou, S.; Gao, R.; Dong, J.; Zhao, X.; Chen, W.; et al. Genesis and Dissemination of Highly Pathogenic H5N6 Avian Influenza Viruses. *J. Virol.* **2017**, *91*. [CrossRef]

18. Yang, H.; Carney, P.J.; Mishin, V.P.; Guo, Z.; Chang, J.C.; Wentworth, D.E.; Gubareva, L.V.; Stevens, J. Molecular Characterizations of Surface Proteins Hemagglutinin and Neuraminidase from Recent H5Nx Avian Influenza Viruses. *J. Virol.* **2016**, *90*, 5770–5784. [CrossRef]

19. Dhingra, M.S.; Artois, J.; Robinson, T.P.; Linard, C.; Chaiban, C.; Xenarios, I.; Engler, R.; Liechti, R.; Kuznetsov, D.; Xiao, X.; et al. Global mapping of highly pathogenic avian influenza H5N1 and H5N× clade 2.3.4.4 viruses with spatial cross-validation. *eLife* **2016**, *5*, e19571. [CrossRef]

20. Adlhoch, C.; Fusaro, A.; Kuiken, T.; Niqueux, É.; Terregino, C.; Staubach, C.; Muñoz Guajardo, I.; Baldinelli, F. Scientific report: Avian influenza overview February–May 2020. *EFSA J.* **2020**, *18*, e06194. [CrossRef]

21. World Organization for Animal Health (OIE). *Update on Highly Pathogenic Avian Influenza in Animals (Type h5 and h7)*; OIE: Paris, France, 2014; Available online: http://www.oie.int/en/animal-health-in-the-world/update-on-avian-influenza/2018/ (accessed on 11 March 2021).

22. Kang, Y.; Liu, L.; Feng, M.; Yuan, R.; Huang, C.; Tan, Y.; Gao, P.; Xiang, D.; Zhao, X.; Li, Y.; et al. Highly pathogenic H5N6 influenza A viruses recovered from wild birds in Guangdong, southern China, 2014–2015. *Sci. Rep.* **2017**, *7*, 44410. [CrossRef]

23. Schrauwen, E.J.; Fouchier, R.A. Host adaptation and transmission of influenza A viruses in mammals. *Emerg. Microbes Infect.* **2014**, *3*, e9. [CrossRef]

24. FAO H7N9. Food and Agriculture Organization of the United Nations. H7N9 Situation Update. Available online: www.fao.org/ag/againfo/programmes/en/empres/H7N9/situation_update.html (accessed on 11 March 2021).

25. Sutton, T.C. The Pandemic Threat of emerging H5 and H7 avian influenza viruses. *Viruses* **2018**, *10*, 461. [CrossRef]

26. Caron, A.; Morand, S.; Garine-Wichatitsky, M.D. Epidemiological interaction at the wildlife/livestock/human interface: Can we anticipate emerging infectious diseases in their hotspots? A framework for understanding emerging diseases processes in their hot spots. *New Front. Mol. Epi Infect. Dis.* **2012**, *2012*, 311–332.

27. Ilyicheva, T.N.; Durymanov, A.G.; Svyatchenko, S.V.; Marchenko, V.Y.; Sobolev, I.A.; Bakulina, A.Y.; Goncharova, N.I.; Kolosova, N.P.; Susloparov, I.M.; Pyankova, O.G.; et al. Humoral immunity to influenza in an at-risk population and severe influenza cases in Russia in 2016–2017. *Arch. Virol.* **2018**, *163*, 2675–2685. [CrossRef] [PubMed]

28. Hoa, L.N.M.; Tuan, N.A.; My, P.H.; Huong, T.T.K.; Chi, N.T.Y.; Hau, T.T.T.; Carrique-Mas, J.; Duong, M.T.; Tho, N.D.; Hoang, N.D.; et al. Assessing evidence for avian-to-human transmission of influenza A/H9N2 virus in rural farming communities in northern Vietnam. *J. Gen. Virol.* **2017**, *98*, 2011–2016. [CrossRef] [PubMed]

29. Bui, V.N.; Nguyen, T.T.; Nguyen-Viet, H.; Bui, A.N.; McCallion, K.A.; Lee, H.S.; Than, S.T.; Coleman, K.K.; Gray, G.C. Bioaerosol Sampling to Detect Avian Influenza Virus in Hanoi's Largest Live Poultry Market. *Clin. Infect. Dis.* **2018**. [CrossRef]

30. Le, T.; Phan, L.; Nguyen, L.; Nguyen, H.; Ly, K.; Ho, T.N.; Trinh, T.; Nguyen, T.M. Fatal Avian Influenza A/H5N1 Infection in a 36-Week Pregnant Woman Survived by her Newborn—Soc Trang, Vietnam, 2012. *Influenza Other Respir Viruses* **2018**. [CrossRef] [PubMed]

31. Thuy, D.M.; Peacock, T.P.; Bich, V.T.N.; Fabrizio, T.; Hoang, D.N.; Tho, N.D.; Diep, N.T.; Nguyen, M.; Hoa, L.N.M.; Trang, H.T.T.; et al. Prevalence and diversity of H9N2 avian influenza in chickens of Northern Vietnam, 2014. *Infect. Genet. Evol.* **2016**, *44*, 530–540. [CrossRef]

32. Mostafa, A.; Abdelwhab, E.M.; Mettenleiter, T.C.; Pleschka, S. Zoonotic potential of influenza A viruses: A comprehensive overview. *Viruses* **2018**, *10*, 497. [CrossRef]

33. Lee, A.C.Y.; Zhu, H.; Zhang, A.J.X.; Li, C.; Wang, P.; Li, C.; Chen, H.; Hung, I.F.N.; To, K.K.W.; Yuen, K.-Y. Suboptimal humoral immune response against influenza A(H7N9) virus is related to its internal genes. *Clin. Vaccine Immunol.* **2015**, *22*, 1235–1243. [CrossRef]

34. Kayali, G.; Setterquist, S.F.; Capuano, A.W.; Myers, K.P.; Gill, J.S.; Gray, G.C. Testing human sera for antibodies against avian influenza viruses: Horse RBC hemagglutination inhibition vs. microneutralization assays. *J. Clin. Virol.* **2008**, *43*, 73–78. [CrossRef] [PubMed]

35. Morens, D.V.; Daszak, P.; Taubenberger, J.K. Escaping Pandora's Box—Another Novel Coronavirus. *N. Engl. J. Med.* **2020**, *382*, 1293–1295. [CrossRef] [PubMed]

36. Taubenberger, J.K.; Kash, J.C.; Morens, D.M. The 1918 influenza pandemic: 100 years of questions answered and unanswered. *Sci. Transl. Med.* **2019**, *11*, eaau5485. [CrossRef]

37. Thông Tin Về Tình Hình Dịch Cúm Gia Cầm, LMLM Và Tai Xanh Ngày 20/02/2017. Available online: http://www.cucthuy.gov.vn/Pages/thong-tin-ve-tinh-hinh-dich-cum-gia-cam-lmlm-va-tai-xanh-ngay-20-02-2017.aspx (accessed on 5 March 2021).

38. Rowe, T.; Abernathy, R.A.; Hu-Primmer, J.; Thompson, W.W.; Lu, X.; Lim, W.; Fukuda, K.; Cox, N.J.; Katz, J.M. Detection of antibody to avian influenza A (H5N1) virus in human serum by using a combination of serologic assays. *J. Clin. Microbiol.* **1999**, *37*, 937–943. [CrossRef] [PubMed]

39. Marchenko, V.; Goncharova, N.; Susloparov, I.; Kolosova, N.; Gudymo, A.; Svyatchenko, S.; Danilenko, A.; Durymanov, A.; Gavrilova, E.; Maksyutov, R.; et al. Isolation and characterization of H5Nx highly pathogenic avian influenza viruses of clade 2.3.4.4 in Russia. *Virology* **2018**, *525*, 216–223. [CrossRef] [PubMed]

40. Susloparov, I.M.; Goncharova, N.; Kolosova, N.; Danilenko, A.; Marchenko, V.; Onkhonova, G.; Evseenko, V.; Gavrilova, E.; Maksutov, R.A.; Ryzhikov, A. Genetic Characterization of Avian Influenza A(H5N6) Virus Clade 2.3.4.4, Russia, 2018. *Emerg. Infect. Dis.* **2019**, *25*, 2338–2339. [CrossRef]
41. WHO. *World Health Organization Surveillance Network: Manual for the Laboratory Diagnosis and Virological Surveillance of Influenza*; World Health Organization: Geneva, Switzerland, 2011.

Article

Seroprevalence of Antibodies to SARS-CoV-2 in Guangdong Province, China between March to June 2020

Cheng Xiao [1,†], Nancy Hiu Lan Leung [2,†], Yating Cheng [3,4,†], Hui Lei [1], Shiman Ling [1], Xia Lin [1], Ran Tao [3,4], Xianzhong Huang [3,4], Wenda Guan [1], Zifeng Yang [1,5], Benjamin John Cowling [2], Mark Zanin [1,2] and Sook-San Wong [1,2,*]

[1] State Key Laboratory of Respiratory Disease, National Clinical Research Center for Respiratory Disease, Guangzhou Institute of Respiratory Health, The First Affiliated Hospital of Guangzhou Medical University, Guangzhou Medical University, Guangzhou 510182, China; xiaocheng1004@stu.gzhmu.edu.cn (C.X.); 2019310641@stu.gzhmu.edu.cn (H.L.); 2019217860@stu.gzhmu.edu.cn (S.L.); linxia@stu.gzhmu.edu.cn (X.L.); guanwenda2004@163.com (W.G.); jeffyah@163.com (Z.Y.); mark.zanin@gird.cn (M.Z.)

[2] WHO Collaborating Centre for Infectious Disease Epidemiology and Control, School of Public Health, Li Ka Shing Faculty of Medicine, The University of Hong Kong, Hong Kong, China; leungnan@hku.hk (N.H.L.L.); bcowling@hku.hk (B.J.C.)

[3] Laboratory Diagnosis Department, Guangzhou Kingmed Center for Clinical Laboratory, International Biotech Island, Guangzhou 510182, China; labcyt@kingmed.com.cn (Y.C.); labtr@kingmed.com.cn (R.T.); labhxz@kingmed.com.cn (X.H.)

[4] Department of Clinical Laboratory Medicine, Guangzhou Medical University, Guangzhou 510182, China

[5] Faculty of Chinese Medicine, Macau University of Science and Technology, Macau 519020, China

[*] Correspondence: sook-san.wong@GIRD.cn; Tel.: +86-178-2584-6078

[†] Authors contributed equally to this work.

Citation: Xiao, C.; Leung, N.H.L.; Cheng, Y.; Lei, H.; Ling, S.; Lin, X.; Tao, R.; Huang, X.; Guan, W.; Yang, Z.; et al. Seroprevalence of Antibodies to SARS-CoV-2 in Guangdong Province, China between March to June 2020. *Pathogens* **2021**, *10*, 1505. https://doi.org/10.3390/pathogens10111505

Academic Editors: Philipp A. Ilinykh and Kai Huang

Received: 13 April 2021
Accepted: 13 May 2021
Published: 18 November 2021

Publisher's Note: MDPI stays neutral with regard to jurisdictional claims in published maps and institutional affiliations.

Abstract: Guangdong province, located in South China, is an important economic hub with a large domestic migrant population and was among the earliest areas to report COVID-19 cases outside of Wuhan. We conducted a cross-sectional, age-stratified serosurvey to determine the seroprevalence of antibodies against SARS-CoV-2 after the emergence of COVID-19 in Guangdong. We tested 14,629 residual serum samples that were submitted for clinical testing from 21 prefectures between March and June 2020 for SARS-CoV-2 antibodies using a magnetic particle based chemiluminescent enzyme immunoassay and validated the results using a pseudovirus neutralization assay. We found 21 samples positive for SARS-CoV-2 IgG, resulting in an estimated age- and sex-weighted seroprevalence of 0.15% (95% CI: 0.06–0.24%). The overall age-specific seroprevalence was 0.07% (95% CI: 0.01–0.24%) in persons up to 9 years old, 0.22% (95% CI: 0.03–0.79%) in persons aged 10–19, 0.16% (95% CI: 0.07–0.33%) in persons aged 20–39, 0.13% (95% CI: 0.03–0.33%) in persons aged 40–59 and 0.18% (95% CI: 0.07–0.40%) in persons ≥60 years old. Fourteen (67%) samples had pseudovirus neutralization titers to S-protein, suggesting most of the IgG-positive samples were true-positives. Seroprevalence of antibodies to SARS-CoV-2 was low, indicating that there were no hidden epidemics during this period. Vaccination is urgently needed to increase population immunity to SARS-CoV-2.

Keywords: SARS-CoV-2; coronavirus disease 2019; seroprevalence; Guangdong; China; antibody

1. Introduction

Guangdong, located in South China, is the most populous province and an important economic hub of China. It has a population of 115 million and the largest economy in China [1]. The major economic and urban areas of Guangdong are centered around nine cities within the Pearl River Delta (PRD) that has a combined population of 63 million people. Due to its economic importance, 30% of its population consists of migrants from other provinces, making it the province with the most domestic migration in China [2]. Shenzhen and Guangzhou, both first-tier megalopolises in Guangdong, are major domestic and international transportation hubs and were among the earliest cities outside of Wuhan to report cases of COVID-19 [3,4]. Guangdong was also among the first provinces in China to declare a Level I public health emergency, the nation's highest emergency response level, on 23 January (Table S1) [5].

Seroprevalence studies have been used to inform the extent of transmission (including subclinical infections), herd immunity and the effectiveness of case-based surveillance in the community [6–8]. Given the importance of Guangdong as an economic, domestic migration and international transport hub, we investigated the community-wide SARS-CoV-2 virus transmission in Guangdong after the emergence of the virus, by conducting a cross-sectional seroprevalence study across province. Our aim was to estimate the age-specific seroprevalence of antibodies to SARS-CoV-2 in Guangdong province after the first wave of COVID-19 in China.

2. Results

2.1. Reported COVID-19 Cases in Guangdong Province between 19 January and 1 July 2020

Between 19 January and 1 July 2020, Guangdong recorded 1961 laboratory-confirmed SARS-CoV-2 infections, with 1642 symptomatic COVID-19 cases, over three waves of activity that were characterized by distinct infection sources (Figure 1). The first wave, from January 14 to February 29, with a total of 1350 COVID-19 cases, was largely seeded by cases originating from Wuhan or Hubei [9] and remained the largest epidemic wave in the province so far. Cases were reported in 20 of the 21 prefecture-level cities in Guangdong (Yunfu reported no case) and were also highest in cities that had high migration activities from Wuhan in the five days prior to the national lockdown (Table 1, Figure 2A). The second outbreak, from March 1 to April 1, involved 161 infections that were mainly imported internationally while the third outbreak from 2 April to 2 May, involved 310 imported and associated-local cases. By 1 July, 1020 (62%) of the COVID-19 cases were reported mainly in Guangzhou and Shenzhen, which were also the major international port-of-entries, while the surrounding cities in the PRD recorded 478 (29%) cases (Table 1). As minimal cases were reported in all other cities after the first wave, the transmission trend across the province during the sera sampling period largely remained the same as in March.

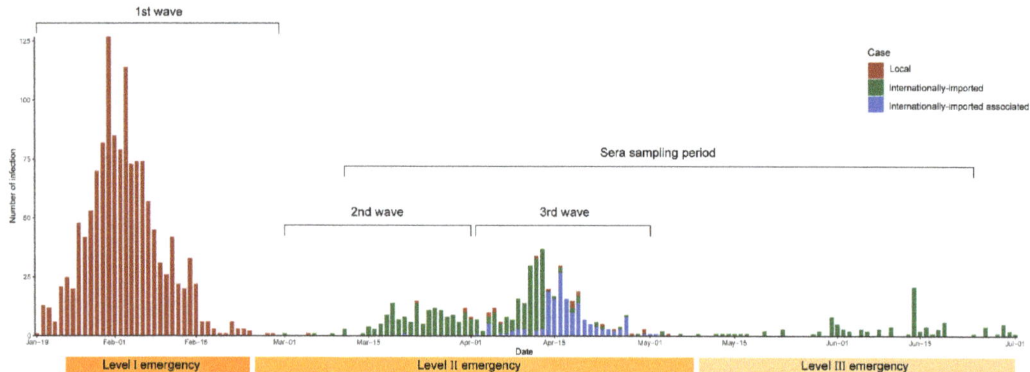

Figure 1. Cases of SARS-CoV-2 infections in Guangdong between 19 January and 1 July 2020. Case numbers represent the total of symptomatic and asymptomatic infections. (Local cases: Cases infected in Guangdong or imported from other provinces in China; Internationally imported cases: Individuals with SARS-CoV-2 infections returned from overseas; Internationally imported associated cases: Local cases identified as being associated with the internationally imported cases.).

2.2. Seroprevalence in Guangdong

Between 11 March and 24 June 2020, 14,629 sera were collected from 983 institutions across Guangdong. A total of 5264 (36%) and 9365 (64%) sera were collected from the low and high-risks cities, respectively. Large cities in high-risks area were generally better sampled while samples in the 10 to 19 years old group were generally under sampled, particularly in the-low risk cities (Table S2).

Out of 14,629 sera, we identified 21 (0.14%) samples positive for SARS-CoV-2 IgG by magnetic particle based chemiluminescent enzyme immunoassay (CLIA). We calculated that in Guangdong province overall, the estimated age and sex-weighted seroprevalence was 0.15% (95% CI 0.06% to 0.24%) (Figure 3A, Table S3). The weighted seroprevalence in high-risk cities was 0.19% (95% CI, 0.06% to 0.33%) (Figure 3B), approximately 2.7-fold higher than the weighted seroprevalence for low-risk cities of 0.07% (95% CI, 0% to 0.24%) (Figure 3C). In the whole of Guangdong, the lowest seroprevalence was detected in the youngest age-group ≤9 years old (0.07% (95%CI, 0.01% to 0.24%), while the seroprevalence estimates in the other age-groups were between 0.13% to 0.22% (Figure 3A). We noted apparent differences in the age-specific trends of seroprevalence estimates between the low and high risks region in Guangdong. In high-risk cities, age-specific seroprevalence was lowest in children ≤9 years of age, highest in adolescents, and lower in the three adult age groups (Figure 3B). In the low-risk cities, age-specific seroprevalence was higher in children 9 years old and older adults ≥40 years old, and lower in younger adults (Figure 3C).

Table 1. Demographic of the twenty-one prefecture-cities in Guangdong; number of confirmed COVID-19 cases between January 19 to March 3, representing the first COVID-19 wave in Guangdong; and between January 19 to July 1, representing the first 6-months post COVID-19 emergence in Guangdong.

Prefecture	Total Cases by March 3 [a] *n*(%)	Total Cases by July 1 [a] *n*(%)	Population (Million) [b]	Female (%) [a]	Incidence (Per Million)	Designated Risk Level	Migration Index with Wuhan [c]
Yunfu	0 (0)	0 (0)	2.5269	49	0	Low	0
Heyuan	4 (0.3)	5 (0.3)	3.0939	49	1.29	Low	0
Jieyang	8 (0.6)	11 (0.7)	6.0894	49	1.32	Low	0
Shanwei	5 (0.4)	5 (0.3)	2.9936	47	1.67	Low	0
Chaozhou	5 (0.4)	6 (0.4)	2.6566	50	1.81	Low	0
Maoming	14 (1.0)	14 (0.9)	6.3132	47	2.22	Low	0
Meizhou	16 (1.2)	17 (1.0)	4.3788	51	2.91	Low	0
Zhanjiang	22 (1.6)	24 (1.5)	7.3320	47	3.00	Low	1,2
Qingyuan	12 (0.9)	12 (0.7)	3.8740	49	3.10	Low	2
Shaoguan	10 (0.7)	10 (0.6)	2.9976	50	3.34	Low	2
Shantou	25 (1.9)	26 (1.6)	5.6385	50	4.44	Low	2
Zhaoqing *	19 (1.4)	20 (1.2)	4.1517	49	4.58	Low	0
Jiangmen *	23 (1.7)	24 (1.5)	4.5982	49	5.01	Low	2
Yangjiang	14 (1.0)	14 (0.9)	2.5556	47	5.51	Low	0
Foshan *	84 (6.2)	100 (6.1)	7.9057	46	10.63	High	2
Dongguan *	99 (7.3)	100 (6.1)	8.3922	44	11.80	High	1,2
Huizhou *	62 (4.6)	62 (3.8)	4.8300	47	12.84	High	1,2
Zhongshan *	66 (4.9)	69 (4.2)	3.3100	46	19.94	High	2
Guangzhou *	346 (25.6)	558 (33.9)	14.9044	49	23.22	High	1,2
Shenzhen *	418 (31.0)	462 (28.2)	13.0266	46	32.10	High	1,2
Zhuhai *	98 (7.3)	103 (6.3)	1.8911	48	51.85	High	1,2
Total (%)	1350 (100)	1642 (100)	113.46				
Average				48	9.65		

[a] Does not include asymptomatic cases. [b] Source: Based on 2018 population data (Guangdong Provincial Bureau of Statistics). [c] 0 = low migration activity with Wuhan, 1 = high incoming migration from Wuhan, 2 = high outgoing migration to Wuhan. All data was for the period between 18 to 22 January 2020. * Prefectural cities in the Pearl River Delta.

In terms of geographical distribution, the seropositivity correlated with the number of reported COVID-19 cases in each city, with the notable exception of Guangzhou. Despite reporting the highest number of reported COVID-19 cases, it had the lowest seropositivity at 0.08% out of 2520 tested samples (Figure 2B, Table S2). This could be due to the high numbers of imported-associated cases that were detected during border screening and were subsequently quarantined at centralized facilities until determined to be PCR-negative, which effectively reduced the risk of virus spreading. In the low-risk region, seropositive samples were detected in Qingyuan (*n* = 1, 0.09%), Jiangmen (*n* = 1, 0.50%) and Shantou (*n* = 3, 0.60%). These three were amongst the low-risk cities in Guangdong that reported cases during the first wave and had high migrant connectivity with Wuhan (Table 1).

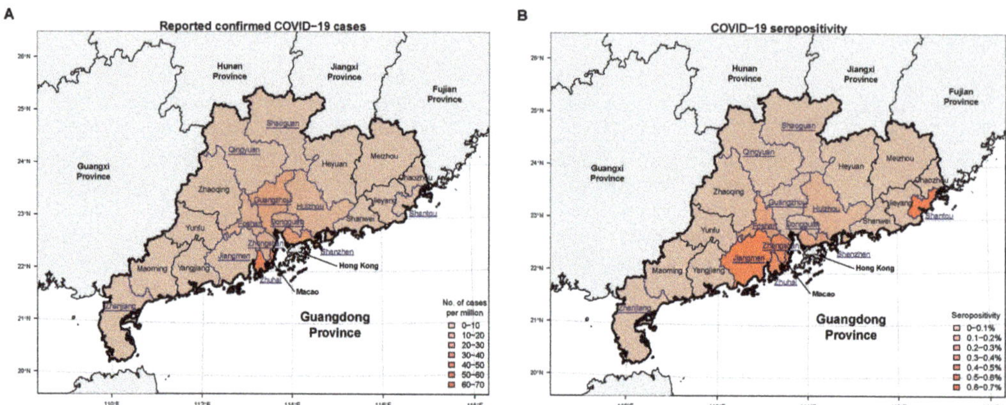

Figure 2. (**A**) Reported confirmed COVID-19 cases based on local official surveillance data in the different prefectural cities in Guangdong province. Asymptomatic infections were not available at a prefectural-city level. (**B**) Seropositivity of antibodies to SARS-CoV-2 as identified from the present study, in prefectural cities of the Guangdong province within the first six-months post COVID-19 emergence. Cities that had relatively higher connectivity with Wuhan prior to 23 January 2020, were underlined and highlighted in blue.

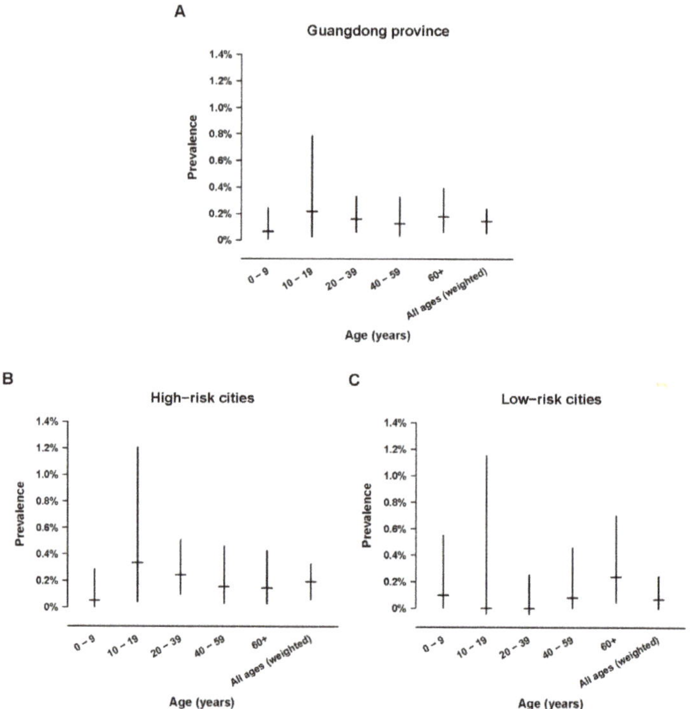

Figure 3. Estimates of age-specific SARS-CoV-2 seroprevalence in Guangdong province in (**A**) all cities or stratified by (**B**) high or (**C**) low risk of COVID-19 activities according to local official surveillance data. Seroprevalences in all ages were age- and sex-weighted according to the population structure of the included cities.

2.3. Proportion of Seropositive Samples with Neutralizing Titers

We tested the SARS-CoV-2 IgG positive samples for neutralization titers with the pseudovirus neutralization (pN) assay using a construct expressing the Spike (S) protein. Of the 21 samples, 14 (67%) had detectable neutralization activity at titers >20 (Table S4). The seven samples that did not have detectable neutralization activity had signal to cut-off readout (S/CO) that ranged from 1.002 to 2.442. There was no significant correlation between IgG-titer (expressed by the S/CO readout) with the measured IC_{50} titer (Pearson's coefficient, r = 0.223, p = 0.33. Figure 4).

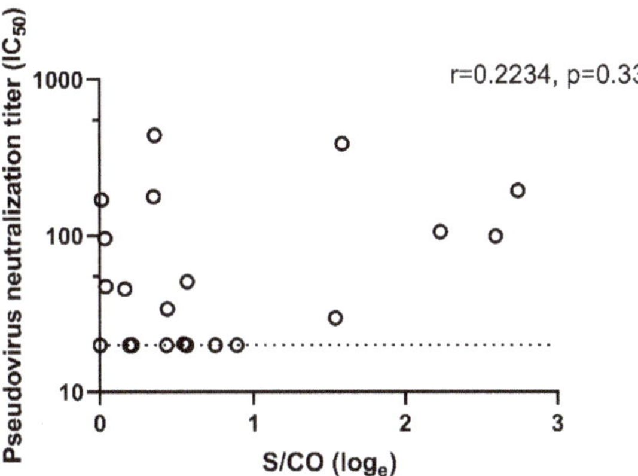

Figure 4. Correlation between the SARS-CoV-2 IgG titers and pseudovirus neutralization titers, expressed as 50% inhibitory concentration (IC_{50}). The signal to cut-off (S/CO) readout on the X-axis were loge transformed to aid visualization of data points.

3. Discussion

In collaboration with a clinical testing laboratory, we were able to use residual serum samples that were submitted for clinical tests to conduct a cross-sectional serosurvey across the expanse of Guangdong province shortly after the emergence of COVID-19 in January 2020. Using a pseudovirus neutralization assay, we confirmed that 67% of the samples had neutralization titers, suggesting that most of the IgG-positive samples were true-positives. The remaining seven samples may still represent true positives as some infections may not have induced neutralizing antibodies, or their neutralizing antibodies may have waned to below the threshold of detection since being infected [10,11]. Studies have shown that the long-term antibody dynamics, particularly in those mild or asymptomatic COVID-19 cases can be variable [12,13]. The lack of correlation between the SAR-CoV-2 IgG-readout measured by CLIA with the pseudovirus neutralization titer could be due to the assay choices. The CLIA detects IgG against both S1 and N protein whereas our pseudovirus in the neutralization assay expressed only the S-protein and would therefore only account for neutralizing activity afforded by S-specific IgG.

Six months into the pandemic, the seroprevalence estimates based on residual sera collected from a clinical diagnostic laboratory network reported in our study were similar to other studies in the general community and lower than studies of high-risk individuals such as healthcare workers or hospital visitors in China (summarized in Table 2). Notably, two studies that included cohorts from Guangdong reported higher seroprevalence rates than ours. For example, by April 2020, Xu et al. found a seroprevalence of about 4% in healthcare workers or their exposed contacts in Wuhan, compared to approximately 1% in healthcare workers or factory workers in Guangzhou [14]. Separately, Liang et al. reported a seroprevalence of 2.1% and 0.6% amongst the 16,000 hospital patients and visitors in Wuhan and Guangzhou between January 25 to April 30 [15]. Collectively, these studies confirmed that SARS-CoV-2 virus transmission in other areas were minimal compared to

Wuhan, but they did not include any validation tests to confirm the antibody specificity and provided limited information with regard to age-specific seroprevalence.

Table 2. Summary of seroprevalence studies of SARS-CoV-2 antibodies in China within the first 6-months of COVID-19 emergence.

Author	Location	Sampling Period	Seropositivity for IgG or IgG and IgM (%)	95% Confidence Interval	Total Population Surveyed	Sampling Population	Approach
Xu et al. [14]	Wuhan	March–April 2020	3.8	2.6 to 5.4	714	Healthcare Workers	Screening
	Guangzhou,		2.80	1.8 to 4.6	563	Hemodialysis Patients	
	Foshan		1.20	0.4 to 3.3	260	Healthcare Workers	
			1.40	0.6 to 2.9	442	Factory Workers	
	Sichuan		0.58	0.45 to 0.76	9442	General Community	
Liang et al. [15]	Wuhan	January–April 2020	2.10	Not reported	8272	Hospital Visitors	Residual Sera
	Guangzhou		0.60	Not reported	8782	Hospital Visitors	Residual Sera
Pan et al. [16]	Wuhan	May 2020	2.39	2.27 to 2.52	61,437	General Community	Cluster, Sampling
Liu et al. [17]	Wuhan	February–April 2020	4.60	4.3 to 4.9	19,555	General Workers	Screening
Chang et al. [18]	Shijiazhuang	April 2020	0.0074	0.0013 to 0.042	13,540	Blood Donors	Residual Sera
	Shenzhen		0.029	0.0081 to 0.11	6810		
	Wuhan		2.29	2.08 to 2.52	17,794		

Our seroprevalence estimate is in line with the results of a large-scale nucleic acid testing conducted by Guangdong Centers for Disease Control and Prevention, which reported low number of PCR-positivity [9,19]. Between January 30 to March 19, they reported 1388 PCR-positive cases out of 1.6 million samples tested (0.089% positivity) [9]. In a follow-up study for the period up to July 9, only 385 samples in over 3.2 million samples tested by third-party institutions, were found to be positive (0.012% positivity) [19]. Notably, they observed changing age trends amongst the positive cases over time. In contrast to the first COVID-19 wave, during which the elderly ($\geq$60 years old) comprised a significant proportion of infected cases (22.2%), PCR-positivity rates in the elderly declined during the subsequent waves (1.9%) but increased in the younger demographic, particularly in those between 20 to 39 years old (59.4%, which increased from 34.4% during the first wave). This was consistent with our data as we did not observe a higher seroprevalence among adults $\geq$60 years old. This trend was in contrast to those observed in Hubei [16], where evidence of infection was more common in those $\geq$60 years old. This suggests that cases in Guangdong after the first wave were more likely amongst the young, mobile travelers, particularly in the high-risk cities.

Guangdong was the epicenter of the first SARS outbreak in 2004 and regularly experienced zoonotic avian influenza cases. Consequently, the province had established an efficient surveillance and response system to emerging pathogens. The rapid response in the province appeared to have been effective in controlling the spread and emergence of SARS-CoV-2 locally. However, our seroprevalence trends suggests that the younger and mobile population in the urban centers of Guangdong as well as smaller cities with high connectivity may be a transmission risk and should be monitored. In conclusion, our data suggests an extremely low seroprevalence across Guangdong.

Limitations of the Study

There were several limitations in our study. One major limitation was that the original sampling design was aimed at evaluating the age-specific seroprevalence in Guangdong at a city-prefecture level. However, the overall low seropositivity precluded this and we instead derived the seroprevalence estimates according to region of epidemic activity. Another limitation was the sampling bias that occurred between urban centers and small cities. It was easier to sample in urban centers due to the larger number of medical institutions available. Sampling was also particularly difficult for persons 10–19 years old, which resulted in the smaller sampling size of younger age groups than that of other age groups in our study. One of the reasons might be that younger individuals were less likely to seek non-emergency medical attention, especially during the period when epidemic control measures were in place. This resulted in wider confidence interval for the estimate of seroprevalence in this age group. Females were slightly oversampled compared to males (Figure S1), and therefore we also weighted by sex in addition to age to provide a more representative estimate. A meta-analysis of >3 millions COVID-19 global cases suggested there is no major difference in the risk of infection between sex [20], and therefore we expect that there would be minimal effect on the overall weighted estimate of seroprevalence due to the oversampling. In addition, our sampling period coincided with the gradual resumption of economic activity within the province as well as the easing of interprovincial travel restrictions. By April 8, the lockdown on Wuhan was lifted, signifying the last major travel restriction within China. As the cities with seropositive samples, including the three cities in the low-risks region, also had high numbers of travelers going to Wuhan in January, we are unable to determine if the positive samples were from a local resident or a returning migrant from Wuhan, or any other province. We also did not conduct IgM-testing, as we were interested in exposure history, for which IgG-antibody titers are more reliable [16]. Finally, as no data were available on asymptomatic infections prior to 1 April 2020 the number of infections in the early days that were reported will likely be underestimated.

4. Materials and Methods

4.1. Epidemiologic Data Source

Since 20 January 2020, COVID-19 has been designated as a notifiable infectious disease in China. Confirmed COVID-19 cases were defined based on the China's National Health Council guideline issued at time of reporting [21–25] and were updated five times between 22 January and 7 March 2020 to keep up with the latest epidemiological and clinical developments [26]. A specific category for asymptomatic cases, which are cases with positive nucleic acid test but without clinical symptoms, were introduced after January 28, but these data were only publicly available after April 1. During the time of sampling for this study, definitions were based on the 6th edition, which was defined as persons with fever or respiratory symptoms and who had etiological, PCR- or serological evidence of infection (File S1).

Data of confirmed COVID-19 and asymptomatic cases were obtained from the Guangdong Provincial Health Authority (GPHA) (File S2) [27]. Number of asymptomatic cases were only reported at the province level (Figure 1) but not at the city level (Table 1). Case data were presented for the following two time periods: between January 19 to March 3, representing the first wave of COVID-19 in China; and time of study conception and up to July 1, coinciding with the week upon completion of sera sampling. For ease of description, we will use the term "infections" to mean symptomatic and asymptomatic SARS-CoV-2 infections in our manuscript.

Population size of each prefecture were sourced from the provincial statistics website and were based on the 2018 population numbers. However, to calculate the age- and sex-adjusted seroprevalence, the prefecture-city population structure were sourced from the 2015 Guangdong One-Percent Population Sample Survey [28], the most recent of such data that was available. As the period before the Wuhan lockdown coincided with the 2020 Spring Festival migration period, when migrants usually return home, we used the migration index sourced from *Baidu Qianxi* (Available online: https://qianxi.baidu.com/ (accessed on 1 May 2020)) in the five-days before 23 January, as an indicator of the degree of connectivity between Wuhan and cities in Guangdong. We selected the window of a 5-day period before the lockdown (on January 23) in our study as this captured the peak (which occurred on January 23) of the inflow/outflow migration from Hubei and Guangdong. The migration index represents the percentage of the daily number of inbound and outbound events by rail, air and road traffic (provided as File S3).

4.2. Study Design

This study was conducted in collaboration with Kingmed Diagnostic Laboratory Services, which provide services across a large network of hospital and medical institutions in China. We collected a total of 14,629 serum samples that remained after being used in the original clinical tests (residual serum) in the following age groups: 0–9, 10–19, 20–39, 40–59 and ≥60 years old. The serum samples, originally collected in 1 to 2 mL standard serum separator tubes, were submitted within 24 h to Kingmed's central laboratory in Guangzhou from all 21 prefecture-level cities in Guangdong province between 11 March and 24 June 2020. After the original clinical tests were done, the residual sera were stored at −20 °C until use in our study. We determined that at least 300 sera samples collected per group would allow the estimation of age-specific seroprevalence within ±1.7%, assuming a minimal level of 0.1% age-specific seroprevalence following a binomial distribution. We classified the prefectural cities in Guangdong as low-risk or high-risk to represent regions that experienced low and high COVID-19 activity in the province based on the GPHA surveillance data. Low-risks were cities that reported less than 10 confirmed COVID-19 cases per million population, while high-risks were cities that reported 10 or more cases per million population by 3 March 2020. We selected sera submitted for blood chemistry tests but excluded those that were submitted for autoimmune or cancer screening. Information about age, sex, source of serum samples and date of sample collection were also retrieved.

4.3. Serologic Assays

We utilized a tiered-testing system to identify SARS-CoV-2 IgG and neutralizing antibodies as an indicator of past exposure to SARS-CoV-2 infection. We first used a commercially available magnetic particle-based chemiluminescent enzyme immunoassay (CLIA) that detects IgG to two SARS-CoV-2 antigens in a single reaction; a peptide to the Spike (S) and the Nucleocapsid (NP) proteins (Bioscience, China). All samples were inactivated at 56 °C for 30 minutes and tested according to the manufacturer's instructions. S/CO ratio > 1.0 was considered positive. Samples were considered to be SARS-CoV-2 IgG-positive if results from two rounds of testing both yielded S/CO > 1.0. The overall assay specificity and sensitivity was 99.2% and 89.6% based on testing of 282 sera from individuals with laboratory-confirmed COVID-19. However, the assay's sensitivity increases to 100% in samples that were collected 17 days after symptom onset [29].

Samples determined to be positive using the CLIA assay were then tested using a lentivirus-based-pseudovirus expressing the SARS-CoV-2 Spike protein neutralization assay as previously described [30].

4.4. Statistical Analysis

We estimated crude age-specific COVID-19 seroprevalences and the confidence intervals in five age-groups (0–9, 10–19, 20–39, 40–59 and ≥60 years) for each prefecture city, the low- and high-risk cities, or Guangdong province using binomial approximation. We also estimated the age- and sex-weighted seroprevalences for low-risk, high-risk cities and Guangdong province using data on the population structure of Guangdong province based on the 2015 Guangdong One-Percent Population Sample Survey. The estimations were conducted with R version 3.6.3. For the neutralization titer, the IC_{50} was calculated using the three-parameter non-linear regression function in Prism 8.0 (GraphPad). The correlation between S/CO readout and neutralization titer were determined using Pearson's Correlation Test, also in Prism 8.0. The Pearson's correlation test was performed under the assumption that the data was sampled from a Gaussian distribution, with the p-value derived from a two-tailed test.

Supplementary Materials: The following are available online at https://www.mdpi.com/article/10.3390/pathogens10111505/s1, Table S1: Timeline and details of the public health response implemented after the emergence of COVID-19 in Guangdong province. Table S2: The seropositivity for antibodies to SARS-CoV-2 in the prefectural cities of Guangdong province identified in the present study between 11 March and 24 June 2020. Since the number of seropositive samples were too low to estimate seroprevalence at the individual city level, percentage of seropositivity were used instead. Table S3: The age-specific SARS-CoV-2 seroprevalence in Guangdong province, and for cities of low or high risk of COVID-19 activities according to the local official surveillance data. The crude seroprevalences for individual age groups were presented, while the overall age- and sex-weighted seroprevalences for cities of low or high risk of COVID-19 activities, and that of Guangdong province, were presented. Table S4: Pseudovirus neutralization titers, expressed as 50% inhibitory concentration (IC_{50}) of the SARS-CoV-2 IgG-positive samples. Figure S1: The population structure in Guangdong Province and sample structure in this study. File S1: Case definition of COVID-19 according to the Protocol on Prevention and Control of Novel Coronavirus Pneumonia (Edition 2 to 7). File S2: Daily case numbers of COVID-19 in Guangdong between 19 January and 1 July 2020. The daily numbers of asymptomatic cases prior to 1 April 2020 were not publicly accessible. File S3: Daily percentage of inbound and outbound events by rail, air and road traffic to and from Wuhan in the five days prior to the lockdown. Only the daily top 100-destination cities were reported by the website. Cities in Guangdong that were listed in daily the top-100 lists were highlighted.

Author Contributions: S.-S.W., B.J.C., N.H.L.L. and M.Z. conceived and designed the study. Y.C., W.G. and Z.Y. supervised the sample collection. Y.C., R.T., C.X., S.L., X.L. and X.H. were responsible for sample and data management as well as CLIA testing. C.X., H.L. and X.L. performed the PPNT testing. C.X., and N.H.L.L. performed the data curation, retrieval and analyses. S.-S.W., B.J.C., N.H.L.L., C.X. and M.Z. wrote the manuscript. S.-S.W., N.H.L.L. and C.X. verified the underlying data. All authors have read and agreed to the published version of the manuscript.

Funding: This work was supported by the Guangzhou Institute of Respiratory Health Open Project (Funds provided by China Evergrande Group)—[Grant number 2020GIRHHMS12], the Project for Prevention and Control of COVID-19 from Guangzhou Institute of Respiratory Health and the High-Level University Talent Construction Program of Guangzhou Medical University to M.Z. and S.-S.W. [Grant numbers: 02-412-B205002-1015022, 02-412-B205002-1015023]. N.H.L.L. and B.J.C. are supported by the Theme-based Research Scheme project no. T11-712/19-N from the University Grants Committee of the Hong Kong Government.

Institutional Review Board Statement: This study was approved by the Institutional Review Board of the First Affiliated Hospital of Guangzhou Medical University (Ethics Number: 2020153).

Informed Consent Statement: The written informed consent was not required as the archived sera samples were used and anonymously handled in this study.

Data Availability Statement: The data available in this study are available on request from the corresponding author. The data are not publicly available due to ethical reasons.

Acknowledgments: We gratefully acknowledge the staff of Kingmed Diagnostics who provided assistance in the study. We also thank Jin-Cun Zhao from the State-Key Laboratory of Respiratory Disease for kindly providing the plasmids and cells for the pseudovirus neutralization assays. We also thank Min Kang and Tie Song, from Guangdong Centers of Disease Control and Prevention for their expert advice. We thank Zhanpeng Jiang for his assistance in making the figures, and Joyce Chen for providing administrative support to the project. We also thank Tingting Liang for collecting the necessary information.

Conflicts of Interest: The authors declare no conflict of interest. The funders had no role in the design of the study; in the collection, analyses, or interpretation of data; in the writing of the manuscript, or in the decision to publish the results.

References

1. Guangdong Provincial Bureau of Statistics. Analysis of Guangdong's Population Development in 2019. Available online: http://stats.gd.gov.cn/tjfx/content/post_2985688.html (accessed on 28 April 2020).
2. Yang, Q.S.; Zhang, H.X.; Mwenda, K.M. County-Scale Destination Migration Attractivity Measurement and Determinants Analysis: A Case Study of Guangdong Province, China. *Sustainability* **2019**, *11*, 362. [CrossRef]
3. Health Commission of Guangdong Province. The National Health Commission Confirmed the First Confirmed Case of Pneumonia Infected by Novel Coronavirus in Guangdong Province. Available online: http://wsjkw.gd.gov.cn/xxgzbdfk/content/post_2880738.html (accessed on 20 January 2020).
4. Guangzhou Municipal Health Commission. Guangzhou Confirmed 2 Cases of Pneumonia with Novel Coronavirus Infection. Available online: http://wjw.gz.gov.cn/ztzl/xxfyyqfk/yqtb/content/post_5643152.html (accessed on 22 January 2020).
5. Health Commission of Guangdong Province. Guangdong Province Decided to Launch a First-Level Response to a Major Public Health Emergency. Available online: http://wsjkw.gd.gov.cn/xxgzbdfk/fkdt/content/post_2879152.html (accessed on 23 January 2020).
6. Tanne, J.H. Covid-19: US cases are greatly underestimated, seroprevalence studies suggest. *BMJ* **2020**, *370*. [CrossRef] [PubMed]

7. Pollan, M.; Perez-Gomez, B.; Pastor-Barriuso, R.; Oteo, J.; Hernan, M.A.; Perez-Olmeda, M.; Sanmartin, J.L.; Fernandez-Garcia, A.; Cruz, I.; de Larrea, F.N.; et al. Prevalence of SARS-CoV-2 in Spain (ENE-COVID): A nationwide, population-based seroepidemiological study. *Lancet* **2020**, *396*, 535–544. [CrossRef]

8. Stringhini, S.; Wisniak, A.; Piumatti, G.; Azman, A.S.; Lauer, S.A.; Baysson, H.; De Ridder, D.; Petrovic, D.; Schrempft, S.; Marcus, K.; et al. Seroprevalence of anti-SARS-CoV-2 IgG antibodies in Geneva, Switzerland (SEROCoV-POP): A population-based study. *Lancet* **2020**, *396*, 313–319. [CrossRef]

9. Lu, J.; du Plessis, L.; Liu, Z.; Hill, V.; Kang, M.; Lin, H.; Sun, J.; Francois, S.; Kraemer, M.U.G.; Faria, N.R.; et al. Genomic Epidemiology of SARS-CoV-2 in Guangdong Province, China. *Cell* **2020**, *181*, 997–1003 e1009. [CrossRef] [PubMed]

10. Chen, Y.; Zuiani, A.; Fischinger, S.; Mullur, J.; Atyeo, C.; Travers, M.; Lelis, F.J.N.; Pullen, K.M.; Martin, H.; Tong, P.; et al. Quick COVID-19 Healers Sustain Anti-SARS-CoV-2 Antibody Production. *Cell* **2020**, *183*, 1496–1507 e1416. [CrossRef] [PubMed]

11. Lei, Q.; Li, Y.; Hou, H.Y.; Wang, F.; Ouyang, Z.Q.; Zhang, Y.; Lai, D.Y.; Jo-Lewis, B.N.; Xu, Z.W.; Zhang, B.; et al. Antibody dynamics to SARS-CoV-2 in asymptomatic COVID-19 infections. *Allergy* **2020**. [CrossRef] [PubMed]

12. Marien, J.; Ceulemans, A.; Michiels, J.; Heyndrickx, L.; Kerkhof, K.; Foque, N.; Widdowson, M.A.; Mortgat, L.; Duysburgh, E.; Desombere, I.; et al. Evaluating SARS-CoV-2 spike and nucleocapsid proteins as targets for antibody detection in severe and mild COVID-19 cases using a Luminex bead-based assay. *J. Virol. Methods* **2021**, *288*. [CrossRef] [PubMed]

13. Wu, F.; Liu, M.; Wang, A. Evaluating the association of clinical characteristics with neutralizing antibody levels in patients who have recovered from mild COVID-19 in Shanghai, China. *JAMA Intern. Med.* **2020**, *180*, 1405. [CrossRef] [PubMed]

14. Xu, X.; Sun, J.; Nie, S.; Li, H.Y.; Kong, Y.Z.; Liang, M.; Hou, J.L.; Huang, X.Z.; Li, D.F.; Ma, T.; et al. Seroprevalence of immunoglobulin M and G antibodies against SARS-CoV-2 in China. *Nat. Med.* **2020**, *76*. [CrossRef] [PubMed]

15. Liang, W.; Lin, Y.; Bi, J.; Li, J.; Liang, Y.; Wong, S.S.; Zanin, M.; Yang, Z.; Li, C.; Zhong, R.; et al. Serosurvey of SARS-CoV-2 among hospital visitors in China. *Cell Res.* **2020**, *30*, 817–818. [CrossRef]

16. Pan, Y.; Li, X.; Yang, G.; Fan, J.; Tang, Y.; Hong, X.; Guo, S.; Li, J.; Yao, D.; Cheng, Z.; et al. Seroprevalence of SARS-CoV-2 immunoglobulin antibodies in Wuhan, China: Part of the city-wide massive testing campaign. *Clin. Microbiol. Infect.* **2020**. [CrossRef] [PubMed]

17. Liu, T.; Wu, S.; Tao, H.; Zeng, G.; Zhou, F.; Guo, F.; Wang, X. Prevalence of IgG Antibodies to SARS-CoV-2 in Wuhan—Implications for the Ability to Produce Long-Lasting Protective Antibodies against SARS-CoV-2. *MedRxiv* **2020**. [CrossRef]

18. Chang, L.; Hou, W.; Zhao, L.; Zhang, Y.; Wang, Y.; Wu, L.; Xu, T.; Wang, L.; Wang, J.; Ma, J.; et al. The Prevalence of Antibodies to SARS-CoV-2 among Blood Donors in China. *Nat. Commun.* **2021**, *12*, 1–10. [CrossRef]

19. Kang, M. *Three Waves of COVID-19 Epidemic and Adaptive Response Measures in Guangdong Province, China*; Guangdong Provincial Center for Disease Control and Prevention: Guangzhou, China, 2020; Manuscript in preparation.

20. Peckham, H.; de Gruijter, N.M.; Raine, C.; Radziszewska, A.; Ciurtin, C.; Wedderburn, L.R.; Rosser, E.C.; Webb, K.; Deakin, C.T. Male sex identified by global COVID-19 meta-analysis as a risk factor for death and ITU admission. *Nat. Commun.* **2020**, *11*. [CrossRef]

21. National Health Commission of the People's Republic of China. Protocol on Prevention and Control of Novel Coronavirus Pneumonia (Edition 5). Available online: http://www.nhc.gov.cn/jkj/s3577/202002/a5d6f7b8c48c451c87dba14889b30147.shtml (accessed on 4 January 2021).

22. National Health Commission of the People's Republic of China. Pneumonia Prevention and Control Plan for Novel Coronavirus Infection (Second Edition). Available online: http://www.nhc.gov.cn/xcs/zhengcwj/202001/c67cfe29ecf1470e8c7fc47d3b751e88.shtml (accessed on 4 January 2021).

23. National Health Commission of the People's Republic of China. Protocol on Prevention and Control of Novel Coronavirus Pneumonia (Edition 6). Available online: http://www.nhc.gov.cn/jkj/s3577/202003/4856d5b0458141fa9f376853224d41d7.shtml (accessed on 4 January 2021).

24. National Health Commission of the People's Republic of China. Protocol on Prevention and Control of Novel Coronavirus Pneumonia (Edition 3). Available online: http://www.nhc.gov.cn/jkj/s7923/202001/470b128513fe46f086d79667db9f76a5.shtml (accessed on 4 January 2021).

25. National Health Commission of the People's Republic of China. Protocol on Prevention and Control of Novel Coronavirus Pneumonia (Edition 4). Available online: http://www.nhc.gov.cn/jkj/s3577/202002/573340613ab243b3a7f61df260551dd4.shtml (accessed on 4 January 2021).

26. Tsang, T.K.; Wu, P.; Lin, Y.; Lau, E.H.Y.; Leung, G.M.; Cowling, B.J. Effect of changing case definitions for COVID-19 on the epidemic curve and transmission parameters in mainland China: A modelling study. *Lancet Public Health* **2020**, *5*, e289–e296. [CrossRef]

27. Health Commission of Guangdong Province. Available online: http://wsjkw.gd.gov.cn/ (accessed on 4 January 2021).

28. Guangdong Provincial Bureau of Statistics. The Data Bulletin of the 1% Population Sample Survey of Guangdong Province in 2015. Available online: http://stats.gd.gov.cn/tjgb/content/post_1430125.html (accessed on 10 May 2016).

29. Long, Q.X.; Liu, B.Z.; Deng, H.J.; Wu, G.C.; Deng, K.; Chen, Y.K.; Liao, P.; Qiu, J.F.; Lin, Y.; Cai, X.F.; et al. Antibody responses to SARS-CoV-2 in patients with COVID-19. *Nat. Med.* **2020**, *26*, 845–848. [CrossRef] [PubMed]

30. Ou, X.; Liu, Y.; Lei, X.; Li, P.; Mi, D.; Ren, L.; Guo, L.; Guo, R.; Chen, T.; Hu, J.; et al. Characterization of spike glycoprotein of SARS-CoV-2 on virus entry and its immune cross-reactivity with SARS-CoV. *Nat. Commun.* **2020**, *11*, 1620. [CrossRef] [PubMed]

*pathogens*

 MDPI

Article

A Cross-Sectional Study of SARS-CoV-2 Seroprevalence between Fall 2020 and February 2021 in Allegheny County, Western Pennsylvania, USA

Lingqing Xu [1,2], Joshua Doyle [1,2], Dominique J. Barbeau [1,2], Valerie Le Sage [3], Alan Wells [4], W. Paul Duprex [2,3], Michael R. Shurin [4,5], Sarah E. Wheeler [4] and Anita K. McElroy [1,2,*]

1   Division of Infectious Diseases, Department of Pediatrics, School of Medicine, University of Pittsburgh, Pittsburgh, PA 15261, USA; lix16@pitt.edu (L.X.); JDOYLE@pitt.edu (J.D.); DJB176@pitt.edu (D.J.B.)
2   Center for Vaccine Research, University of Pittsburgh, Pittsburgh, PA 15261, USA; pduprex@pitt.edu
3   Department of Microbiology and Molecular Genetics, University of Pittsburgh, Pittsburgh, PA 15261, USA; valerie.lesage@pitt.edu
4   Department of Pathology, School of Medicine, University of Pittsburgh, Pittsburgh, PA 15261, USA; wellsa@upmc.edu (A.W.); shurinmr@upmc.edu (M.R.S.); Wheelerse3@upmc.edu (S.E.W.)
5   Department of Immunology, University of Pittsburgh, Pittsburgh, PA 15261, USA
*   Correspondence: mcelroya@pitt.edu; Tel.: +1-412-648-4026

Citation: Xu, L.; Doyle, J.; Barbeau, D.J.; Le Sage, V.; Wells, A.; Duprex, W.P.; Shurin, M.R.; Wheeler, S.E.; McElroy, A.K. A Cross-Sectional Study of SARS-CoV-2 Seroprevalence between Fall 2020 and February 2021 in Allegheny County, Western Pennsylvania, USA. *Pathogens* **2021**, *10*, 710. https://doi.org/10.3390/pathogens10060710

Academic Editors: Philipp A. Ilinykh and Kai Huang

Received: 10 May 2021
Accepted: 4 June 2021
Published: 6 June 2021

Publisher's Note: MDPI stays neutral with regard to jurisdictional claims in published maps and institutional affiliations.

Abstract: Seroprevalence studies are important for understanding the dynamics of local virus transmission and evaluating community immunity. To assess the seroprevalence for SARS-CoV-2 in Allegheny County, an urban/suburban county in Western PA, 393 human blood samples collected in Fall 2020 and February 2021 were examined for spike protein receptor-binding domain (RBD) and nucleocapsid protein (N) antibodies. All RBD-positive samples were evaluated for virus-specific neutralization activity. Our results showed a seroprevalence of 5.5% by RBD ELISA, 4.5% by N ELISA, and 2.5% for both in Fall 2020, which increased to 24.7% by RBD ELISA, 14.9% by N ELISA, and 12.9% for both in February 2021. Neutralization titer was significantly correlated with RBD titer but not with N titer. Using these two assays, we were able to distinguish infected from vaccinated individuals. In the February cohort, higher median income and white race were associated with serological findings consistent with vaccination. This study demonstrates a 4.5-fold increase in SARS-CoV-2 seroprevalence from Fall 2020 to February 2021 in Allegheny County, PA, due to increased incidence of both natural disease and vaccination. Future seroprevalence studies will need to include the effect of vaccination on assay results and incorporate non-vaccine antigens in serological assessments.

Keywords: SARS-CoV-2; seroprevalence; ELISA; neutralization assay

1. Introduction

Severe Acute Respiratory Syndrome Coronavirus 2 (SARS-CoV-2) spread quickly and caused a worldwide pandemic, affecting social and economic life globally [1]. Diagnosis of SARS-CoV-2 acute infection relies on viral tests such as PCR or virus antigen detection. However, these tests lack the ability to identify prior infections. In contrast, serological assays such as enzyme-linked immunosorbent assays (ELISAs) measure antibody responses to specific virus antigens and are useful for determining the prevalence of a disease in an affected area and can identify individuals as potential donors for convalescent plasma therapeutics [2].

All current Emergency Use Authorization (EUA)-authorized serological tests for SARS-CoV-2 target the nucleocapsid (N) or spike (S) protein. N protein facilitates the replication of viral RNA and the assembly and release of viral particles after infection [3]. S protein binds to the angiotensin-converting enzyme 2 (ACE2) receptor on the surface of human cells for cell entry of SARS-CoV-2 [4,5]. The receptor-binding domain (RBD) of S protein is a main target of anti-viral antibodies [6]. Both S and N proteins are highly expressed during infection and are immunogenic [6,7]. Two leading mRNA-based SARS-CoV-2 vaccines, one developed by Moderna and the other by Pfizer and BioNTech, both use S protein as an immunogen [8,9].

Population-based seroprevalence studies of COVID-19 carried out in hotspots of COVID-19 across the world between March and June 2020 showed a 4.41% seroprevalence in the US and a 3.38% seroprevalence worldwide [10–12]. One nationwide study conducted between July and September 2020 showed a range of 1–23% jurisdiction-level seroprevalence and an estimate of fewer than 10% people with detectable SARS-CoV-2 antibodies, indicating that the majority of the US population had not yet been exposed [13]. Although evidence-based information about the efficacy of COVID-19 interventions is urgently needed in all communities, most studies so far have focused on large scale populations either nationwide or in metropolitan areas and less is known about seroprevalence in medium-sized cities. Freeman et al. reported a seroprevalence of 1% in the first half of 2020 in the immunocompromised pediatric patients in one pediatric quaternary care center in Pittsburgh, PA [14]. However, data for the general population in this area are lacking.

The goal of this study was to develop a serological testing strategy to estimate the seroprevalence of SARS-CoV-2 in the population of Allegheny County, Western PA, in Fall 2020 and February 2021, which are two critical time points before and after the large wave of cases that occurred between December 2020 and January 2021, as well as the introduction of two EUA-approved vaccines in US in Dec 2020. In addition, we used two ELISA assays that enabled us to distinguish infected from vaccinated individuals (DIVA), which allowed for a further comparison of demographics between infected and vaccinated individuals. Overall, we observed a 4.5-fold increase in SARS-CoV-2 seroprevalence from Fall 2020 to February 2021 in Allegheny County, PA, by RBD ELISA. Among samples positive for RBD, 36.4% from Fall 2020 and 66.7% from February 2021 were also positive by neutralization assay. These changes were driven by both natural disease acquisition and vaccination rollout. Additionally, our data showed income and race disparities in infection and vaccination, respectively, suggesting the need to better support people of disadvantaged groups during the pandemic.

2. Results

2.1. SARS-CoV-2 RBD and N ELISA Validation

RBD and N ELISA assays had a limit of detection of 1:100. Specificity of the SARS-CoV-2 ELISA assays was determined using 183 human serum samples collected before the pandemic [15]. This included healthy volunteers, as well as patients who tested positive for other viral or inflammatory conditions, including other human coronaviruses (See Supplementary Table S1. Legend for details).

With an endpoint titer cutoff of 300, the assays had 89.6% (RBD) or 94.5% (N) specificity. With an endpoint titer cutoff of 900, the assays had 97.8% (RBD) or 98.9% (N) specificity (Figure 1a,b). Sensitivity of the assays was determined using 134 human serum samples from patients known to have COVID-19 between 16 March 2020 and 12 May 2020 [15]. In some cases, multiple samples from an individual patient were available on different days post symptom onset. Samples collected on or after Day 14 post symptom onset were used for sensitivity analysis. In this cohort, there was 100% (RBD) and 93.5% (N) sensitivity at an endpoint titer cutoff of 300, and 94.8% (RBD) and 83.1% (N) sensitivity at an endpoint titer cutoff of 900 (Figure 1c,d). An endpoint titer cutoff of 900 was chosen for use in the RBD assay and an endpoint titer cutoff of 300 was chosen for use in the N assay, with a specificity at 97.8% for RBD and 94.5% for N, and a sensitivity at 94.8% for RBD and 93.5% for N. RBD and N ELISAs were further validated using WHO international standards (Supplementary Figure S1).

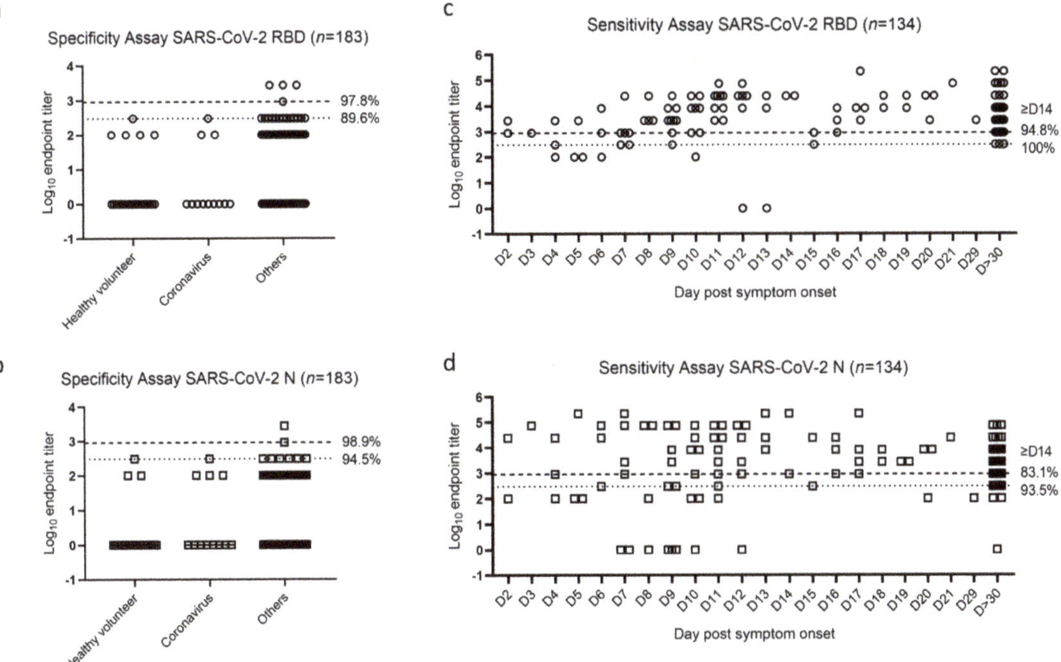

Figure 1. Specificity and sensitivity analysis of RBD and N ELISAs. Each circle represents one sample for RBD (**a,c**) and each square represents one sample for N (**b,d**) in each assay. Samples were grouped by positivity for a specific infection or assay (**a,b**) and see Supplemental Table S1 for details of "Others". Samples from patients with COVID-19 disease are grouped by the day post-self-reported-symptom onset (**c,d**). Dashed line indicates a titer at 900 and dotted line indicates a titer at 300. Numbers next to each line represents the specificity or sensitivity of the assay at the given cutoff value. Samples collected on or after Day 14 post symptom onset were used for sensitivity analysis.

2.2. Seroprevalence of COVID-19 in Allegheny County, Western PA

A total of 199 human blood samples were collected in the Fall of 2020 and 194 samples in February of 2021. A total of 88.5% of the study subjects were from Allegheny County and 9.9% were from other counties in PA (Supplementary Table S2). Women, seniors, and African Americans were more represented in the study cohort compared to that in the population of Allegheny County (Supplementary Table S2).

The endpoint titers for both RBD and N were determined for each sample and using predefined cutoffs, the seroprevalence of SARS-CoV-2 in the Fall cohort was 5.5% by RBD ELISA, 4.5% by N ELISA, and 2.5% for both. In the February cohort, the seroprevalence increased to 24.7% by RBD ELISA, 14.9% by N ELISA, and 12.9% for both (Table 1).

The 11 samples positive for RBD in Fall 2020 included two documented recovered cases of COVID-19 (one male aged 50–59 and one female aged 80+) and two volunteers who were enrolled in the Phase III clinical trial of the Moderna vaccine (Figure 2a).

The remaining seven samples positive for RBD in Fall 2020 were all from young adults aged 19–29, several of which had reported a history of either contact with COVID-19 confirmed cases, or mild COVID-19 related symptoms, but were never tested. This suggested a higher local prevalence of COVID-19 in young adults in Fall 2020 (Figure 2a).

Table 1. Analysis and comparison of seroprevalence and antibody neutralization between Fall 2020 and February 2021.

	Fall 2020 (*n* = 199)	**February 2021 (*n* = 194)**
RBD positive (endpoint titer $\geq$ 900)—No. (%) [95%CI]	11 (5.5) [3–10]	48 (24.7) [19–31]
N positive (endpoint titer $\geq$ 300)—No. (%) [95%CI]	9 (4.5) [2–8]	29 (14.9) [10–21]
RBD and N both positive—No. (%) [95%CI]	5 (2.5) [1–6]	25 (12.9) [9–18]
Neut positive (FRNT$_{50}$ $\geq$ 40)—No. (%) [95%CI]	4 (2.0) [1–5]	32 (16.5) [12–22]

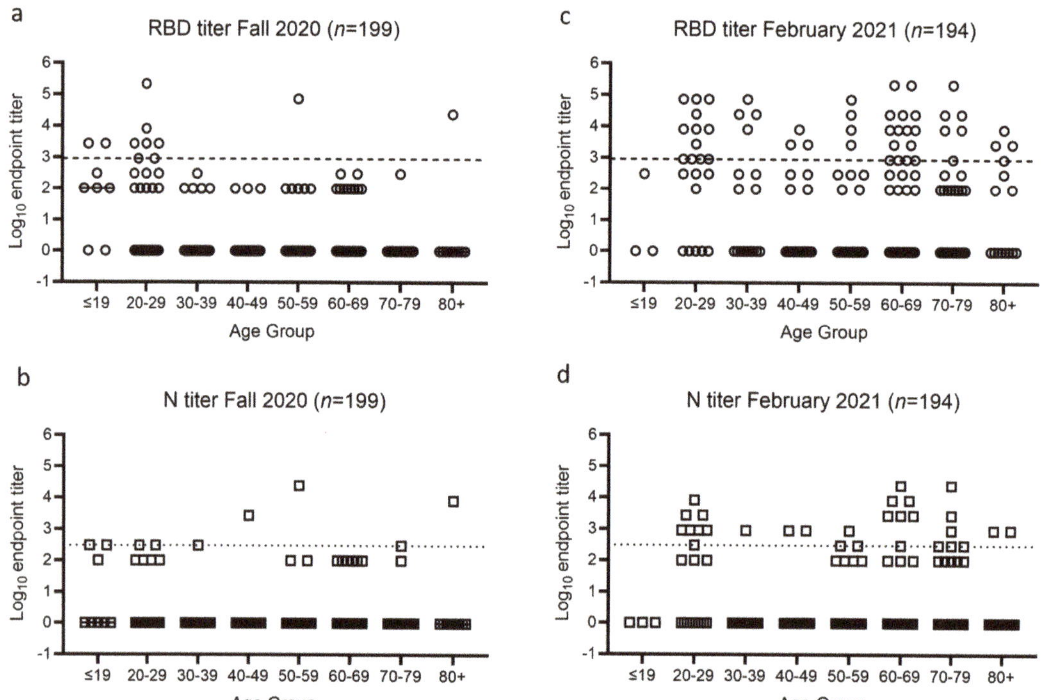

Figure 2. RBD and N endpoint titers of all samples. Each circle represents one sample for RBD (**a,c**) and each square represents one sample for N (**b,d**) at the two time points of the study. Dashed line indicates a titer at 900 as cutoff for RBD positive and dotted line indicates a titer at 300 as cutoff for N positive. Samples are grouped by age.

In the February cohort, the increased seroprevalence suggested increases in either acquisition of natural disease during the winter peak or vaccination following EUA for BNT162b2 by Pfizer and BioNTech on 11 December 2020 and mRNA-1273 by Moderna on 18 December 2020 (Table 1). Both BNT162b2 and mRNA-1273 are S protein mRNA-based vaccines, so vaccination would be expected to elicit RBD antibodies but not N antibodies. In contrast, infection by SARS-CoV-2 would trigger immune responses against both RBD and N. Therefore, based on their RBD and N titers, all 59 individuals who tested positive for RBD were separated into three groups, infected (*n* = 30, RBD+ and N+), vaccinated (*n* = 19, RBD+ and N−), or unclear (*n* = 10 RBD+ and N−) (Figure 3). These classification groups were informed by self-reporting or chart review. The group designated as unclear had RBD titers that were either at or one dilution above the endpoint titer cutoff value and had no corroborating data from chart review or self-reporting.

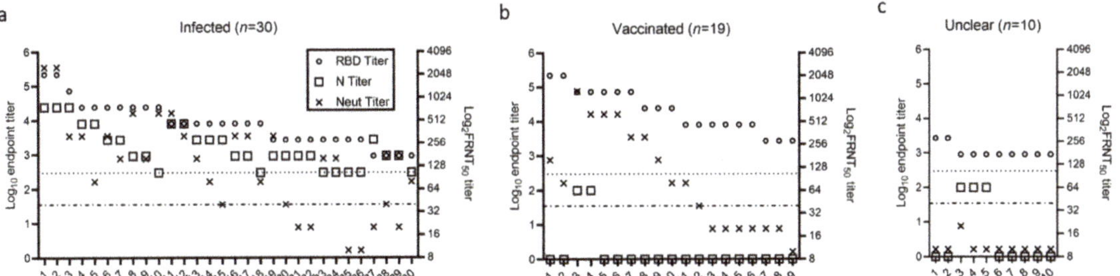

Figure 3. Titer comparison between RBD ELISA, N ELISA, and Neutralization assay. RBD endpoint titer, N endpoint titer, and FRNT$_{50}$ titer of all RBD positive samples in groups of infected (**a**), vaccinated (**b**), and unclear (**c**). Left y axis is Log$_{10}$ endpoint titers for RBD and N ELISAs. Right y axis is Log$_2$FRNT$_{50}$ titer. Dotted line indicates a titer at 300 which was the cutoff titer for N positive. Dash-dotted line indicates a titer of 40 as cutoff for positive neutralization.

All RBD-positive samples were tested in an FRNT$_{50}$ assay. The neutralization assay had a limit of detection of 1:20, and samples were considered positive if the titer was $\geq$40. Among the 30 individuals classified as infected, 24 (80%) were positive by FRNT$_{50}$; the 6 that were unable to neutralize had an RBD titer $\leq$2700 (Figure 3a). In comparison, among the 19 individuals classified as vaccinated, 12 (63.2%) were positive by FRNT$_{50}$; the 7 that were unable to neutralize had an RBD titer $\leq$8100 (Figure 3b). All unclear samples failed to neutralize the virus, even though they had a positive RBD titer (Figure 3c). Comparison between FRNT$_{50}$ and ELISA titers revealed a significant correlation for RBD but not N (Figure 4a,b). Notably, samples with RBD titers at 8100 and positive N titers were more often able to neutralize SARS-CoV-2, than samples with RBD titers at 8100 but negative N titers, suggesting that despite having the same RBD titer, there might be a qualitative difference in spike antibodies generated during infection versus vaccination.

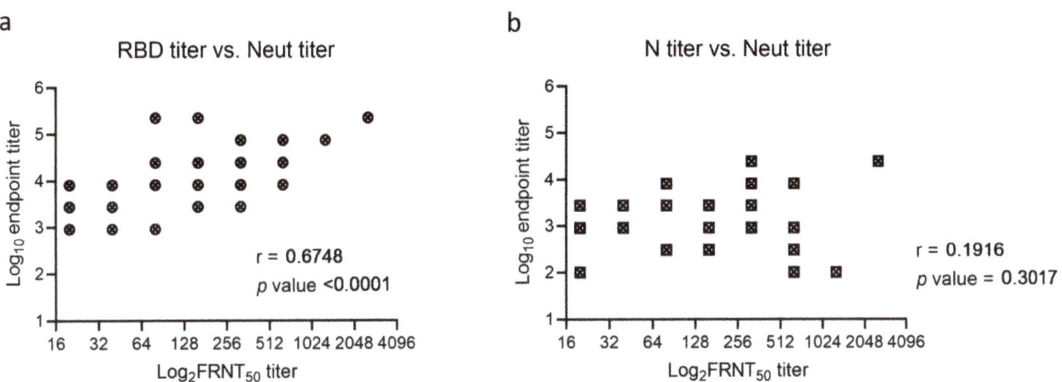

Figure 4. Correlation between ELISA titer and Neutralization titer. Samples from Figure 3 with titers above the detection threshold (100 for RBD and N, 20 for FRNT$_{50}$) of each assay were selected for the correlation analysis between neutralization titer and RBD titer (**a**) or N titer (**b**). Spearman's Rank Correlation Coefficient r and Probability (p) value (two-tailed) are shown.

2.3. Comparison of Income and Race between Infected and Vaccinated Groups

Demographic comparisons were evaluated for age, income, and race in infected (*n* = 25) and vaccinated (*n* = 17) individuals from the February 2021 cohort. In both groups, the average age was approximately 50. The median household income of participants was significantly lower in the infected group than that of the vaccinated group (Supplementary Figure S2a). Three students were removed from this comparison since their income data were felt to be reflective of their guardians' income as their zip codes were out of state. The race distribution between infected and vaccinated groups in the February cohort revealed similar percentages of African American and White in the infected group but a lower percentage of African American in the vaccinated group (Supplementary Figure S2b). These data are consistent with other reports demonstrating the disproportionate effects of the pandemic and vaccine accessibility on lower-income and African American groups in PA as well as nationwide US [16–19].

3. Discussion

The ELISAs used in this study were shown to have a high degree of sensitivity and specificity. These assays are semi-quantitative while most commercially available antibody detection kits only offer qualitative results. The ability to perform functional, neutralizing assays of antibody activity remains limited to BSL-3 laboratory settings. Therefore, an assay such as the RBD ELISA, which is significantly correlated with neutralization titer, has utility for measuring virus-specific activity and could possibly be used to define a correlate of immune protection in clinical and vaccine studies. We observed a 4.5-fold increase in community immunity by RBD ELISA between Fall of 2020 and February of 2021 and were able to demonstrate that this increase was driven both by infection and vaccination.

The correlations between RBD, N, and neutralization were particularly interesting. When a serum sample had a high RBD titer ($\geq$24,300), which indicated strong immune responses elicited by either infection or vaccination, that sample had the ability to neutralize the virus regardless of the N titer. Out of 12 samples with RBD titers at 900, only 2 were positive in the neutralization assay, both of which also had a positive N titer. Two samples with negative RBD titers but positive N titers (one at 900 and the other at 2700) were evaluated for neutralization and neither of them neutralized SARS-CoV-2 virus, suggesting these were false positives or possibly they represent cross reactivity with another coronavirus [20]. When a sample had an RBD titer between 900 and 24,300, it appeared more likely to neutralize the virus if it also had a positive N titer. A total of 12 out of 16 samples from infected individuals with an RBD titer at 2700 or 8100 demonstrated neutralization, whereas only 2 out of 8 samples from vaccinated individuals did. This could have several implications. First, the fact that 4 out of 16 samples with positive RBD and N titers failed to neutralize the virus was consistent with previous findings that people who had contracted COVID-19 could sometimes be reinfected [21,22]. Secondly, the complete vaccination history was not available for all samples, therefore it is possible that some vaccinated cases were not fully vaccinated, meaning that they either received one dose or received the second dose within 14 days at the time of sample collection. Lastly, two infected samples with RBD titer = 2700 and N titer = 300 demonstrated high neutralization titer at 160, suggesting that other factors may also play a role in the generation of neutralizing activity. For example, binding sites on the ACE2 receptor and the binding affinity of an antibody have both been shown to be critical for neutralization potential [23–25]. The SARS-CoV-2 immune response is different in infected versus vaccinated individuals [26]. This could result in a greater breath or alter the class switching and/or affinity maturation of S-specific antibodies in infected versus vaccinated individuals and is consistent with our observations.

The entire population was naïve to SARS-CoV-2 infection when this virus emerged. However, the pandemic has affected people from diverse social and economic backgrounds differently. By comparing income and race distribution between infected and vaccinated groups, we found that people with lower incomes were more likely to have been infected than vaccinated and people of color were less likely to have been vaccinated at the time of sampling. The February 2021 cohort of specimens were obtained when PA restricted vaccination to health care workers, residents/care staff in skilled nursing facilities, individuals ages 65 and older and those ages 16–64 with certain underlying medical conditions [27]. Restrictions on access to vaccination could have influenced these results.

This study has several limitations. Although people of both genders and most age groups and races were included, it had more women, seniors (aged 60–69 and 70–79), and African Americans as compared to the representative distribution in Allegheny County, PA (Supplementary Table S2). In addition, whereas children under 15 make up 15.6% of the population, this study was focused on adults. Finally, patient information regarding previous COVID-19 diagnosis or vaccination was limited to data accessible by chart abstraction for the majority of participants. A major strength of this study is the use of two assays that permits classification of individuals as either vaccinated or infected. With many of the vaccine platforms containing the S antigen, the need for a non-vaccine antigen for community serosurveys of infection is of increased importance. In addition, estimating local seroprevalence using a small cohort of samples can be a valuable tool for quickly assessing the degree of immunity in a community and can inform public health decisions during a pandemic.

4. Materials and Methods

4.1. Human Samples

Human subjects research was performed according to University of Pittsburgh approved IRB protocols (20030228 and 20040220). Positive and negative control samples used for specificity and sensitivity assays were from residual samples at UPMC clinical laboratories and have been previously described [15]. WHO international standards were from the National Institute for Biological Standards and Control (NIBSC). Serum or plasma was collected and stored at $-20\,^{\circ}$C.

Blood samples obtained in Fall 2020 were from two cohorts. One cohort included 100 samples obtained between 21 September 2020 and 13 October 2020 from healthy adults who were willing to have phlebotomy performed for infectious disease research. All participants self-reported their age, sex, race, ethnicity, residential zip code, history of travel and immunization, as well as history of known COVID-19 disease or contact with COVID-19 confirmed cases. A second cohort of 99 samples were residual specimens obtained from the UPMC Mercy clinical laboratory between 2 November 2020 to 4 November 2020; these specimens were from outpatients undergoing routine bloodwork. Blood samples from February 2021 were all residual specimens obtained from the UPMC Mercy outpatient laboratory between 1 February 2021 and 5 February 2021. Age, sex, race, residential zip code and any history of COVID-19 disease or testing was abstracted from the medical record prior to de-identification for all UPMC Mercy outpatient samples.

4.2. Enzyme-Linked Immunosorbent Assay (ELISA)

MaxiSorp™ 96-well plates (Thermofisher) were coated with SARS-CoV-2 RBD-His or SARS-CoV-2 N-His protein (see Supplementary Material for details on clone construction and protein purification) at 50 ng per well diluted in PBS and incubated at $4\,^{\circ}$C overnight. Following removal of coating solution, plates were blocked with blocking buffer (5% non-fat milk in PBS with 0.1% Tween-20, PBST) and incubated at $37\,^{\circ}$C for 1 h. Three-fold serial dilutions of samples were prepared in blocking buffer, then incubated on plates at $37\,^{\circ}$C for 2 h. Plates were washed three times with PBST followed by incubation with donkey anti-human IgG horseradish peroxidase (HRP) conjugated secondary antibody (Jackson ImmunoResearch) diluted 1:10,000 in blocking buffer at $37\,^{\circ}$C for 1 h. Plates were washed again prior to the addition of TMB peroxidase substrate mix (Seracare) and incubated at room temperature (RT) for 5 min. TMB stop solution (Seracare) was added and the optical density (OD) at 450 nm was measured using a Molecular Devices SpectraMax 340PC Microplate Reader. A normal human serum non-reactive to SARS-CoV-2 was included as negative control and a human serum known to be reactive to SARS-CoV-2 was included as positive control on each plate.

4.3. Focus Reduction Neutralization Test (FRNT)

Neutralization assays were performed as previously described [28] except with 2×10^3 FFU/mL of SARS-CoV-2 (University of Pittsburgh clinical isolate from June 2020). Fixation was performed using 4% paraformaldehyde for 20 min at RT. Rabbit anti-SARS-CoV-2 N antibody (GenScript custom, 1:3000 diluted in blocking buffer) was used as primary antibody and goat anti-rabbit IgG HRP (Jackson ImmunoResearch, 1:1000 diluted in blocking buffer) was used as secondary antibody. MossBio TMB substrate was used for foci development. The $FRNT_{50}$ is the dilution of sera that neutralized at least 50% of the input virus.

4.4. Data Analysis

The US Census 2019 ACS 5-year estimates of the population in Allegheny County, PA, was accessed online and data regarding race, age, sex and median household income based upon zip code was abstracted. All graphs were made using GraphPad Prism Version 9. Correlation coefficient (r) and two-tailed probability (p) values were calculated using Spearman's correlation. Confidence intervals were calculated using the Clopper–Pearson exact method. The endpoint titer was defined as the dilution of the serum that gave an OD value at least three standard deviations above the average value obtained from the negative control serum.

Supplementary Materials: The following are available online at https://www.mdpi.com/article/10.3390/pathogens10060710/s1, Figure S1: Comparison of RBD and N titers between the in-house assays and that reported by NIBSC using WHO international standards. Five dots shown on each graph represent samples of one healthy donor collected before 2019 and four COVID-19 recovered patients. Spearman's rank correlation coefficient r and probability (*p*) value (two-tailed) are shown. Figure S2: Comparison of income and race between infected and vaccinated cases in February 2021. Median income distribution in infected and vaccinated groups (a): Y axis is the median household income of the area based on the zip code of the home address reported by each study subject. The line in each group represents the median. * represents a *p* value < 0.05 by an unpaired *t*-test (two-tailed). Race distribution in infected and vaccinated groups (b): Total samples refer to all samples collected in February 2021. Data were compared to that of American Community Survey (ACS) 5-year estimates for Allegheny County, PA, generated by United States Census Bureau. Table S1: Detailed groups of samples used in the Specificity Assays of SARS-CoV-2 RBD and N. Table S2: Demographics of study subjects.

Author Contributions: Conceptualization, A.K.M.; validation, L.X.; formal analysis, L.X.; J.D.; investigation, L.X.; J.D.; resources, A.K.M., V.L.S.; W.P.D.; M.R.S.; S.E.W.; writing—original draft preparation, L.X.; writing—review and editing, A.K.M.; visualization, L.X.; supervision, A.W.; W.P.D.; A.K.M.; project administration, D.J.B.; funding acquisition, A.K.M. All authors have read and agree to the published version of the manuscript.

Funding: This research was funded by the DSF Charitable Foundation, the University of Pittsburgh Clinical and Translational Science Institute (CTSI), the Burroughs Wellcome (CAMS 1013362.02) and Stanley and Susan Plotkin/Sanofi Pasteur Fellowship Award.

Institutional Review Board Statement: The study was conducted according to the guidelines of the Declaration of Helsinki, and approved by the Institutional Review Board of The University of Pittsburgh.

Informed Consent Statement: Informed consent was obtained from all subjects involved in study 20030228. Patient consent was waived for study 20040220 since only residual, deidentified samples were used.

Data Availability Statement: Data is contained within the article or supplementary material.

Conflicts of Interest: The authors declare no conflict of interest.

References

1. Nicola, M.; Alsafi, Z.; Sohrabi, C.; Kerwan, A.; Al-Jabir, A.; Iosifidis, C.; Agha, M.; Agha, R. The socio-economic implications of the coronavirus pandemic (COVID-19): A review. *Int. J. Surg.* **2020**, *78*, 185–193. [CrossRef]
2. Amanat, F.; Stadlbauer, D.; Strohmeier, S.; Nguyen, T.H.O.; Chromikova, V.; McMahon, M.; Jiang, K.; Arunkumar, G.A.; Jurczyszak, D.; Polanco, J.; et al. A serological assay to detect SARS-CoV-2 seroconversion in humans. *Nat. Med.* **2020**, *26*, 1033–1036. [CrossRef] [PubMed]
3. Narayanan, K.; Chen, C.J.; Maeda, J.; Makino, S. Nucleocapsid-independent specific viral RNA packaging via viral envelope protein and viral RNA signal. *J. Virol.* **2003**, *77*, 2922–2927. [CrossRef] [PubMed]
4. Letko, M.; Marzi, A.; Munster, V. Functional assessment of cell entry and receptor usage for SARS-CoV-2 and other lineage B betacoronaviruses. *Nat. Microbiol.* **2020**, *5*, 562–569. [CrossRef] [PubMed]
5. Wells, H.L.; Letko, M.; Lasso, G.; Ssebide, B.; Nziza, J.; Byarugaba, D.K.; Navarrete-Macias, I.; Liang, E.; Cranfield, M.; Han, B.A.; et al. The evolutionary history of ACE2 usage within the coronavirus subgenus Sarbecovirus. *Virus Evol.* **2020**. [CrossRef]
6. Berry, J.D.; Hay, K.; Rini, J.M.; Yu, M.; Wang, L.; Plummer, F.A.; Corbett, C.R.; Andonov, A. Neutralizing epitopes of the SARS-CoV S-protein cluster independent of repertoire, antigen structure or mAb technology. *MAbs* **2010**, *2*, 53–66. [CrossRef]
7. Liu, S.J.; Leng, C.H.; Lien, S.P.; Chi, H.Y.; Huang, C.Y.; Lin, C.L.; Lian, W.C.; Chen, C.J.; Hsieh, S.L.; Chong, P. Immunological characterizations of the nucleocapsid protein based SARS vaccine candidates. *Vaccine* **2006**, *24*, 3100–3108. [CrossRef] [PubMed]
8. Jackson, L.A.; Anderson, E.J.; Rouphael, N.G.; Roberts, P.C.; Makhene, M.; Coler, R.N.; McCullough, M.P.; Chappell, J.D.; Denison, M.R.; Stevens, L.J.; et al. An mRNA vaccine against SARS-CoV-2-preliminary report. *N. Engl. J. Med.* **2020**, *383*, 1920–1931. [CrossRef]
9. Mulligan, M.J.; Lyke, K.E.; Kitchin, N.; Absalon, J.; Gurtman, A.; Lockhart, S.; Neuzil, K.; Raabe, V.; Bailey, R.; Swanson, K.A.; et al. Phase I/II study of COVID-19 RNA vaccine BNT162b1 in adults. *Nature* **2020**, *586*, 589–593. [CrossRef]
10. Eckerle, I.; Meyer, B. SARS-CoV-2 seroprevalence in COVID-19 hotspots. *Lancet* **2020**, *396*, 514–515. [CrossRef]
11. Lai, C.C.; Wang, J.H.; Hsueh, P.R. Population-based seroprevalence surveys of anti-SARS-CoV-2 antibody: An up-to-date review. *Int. J. Infect. Dis.* **2020**, *101*, 314–322. [CrossRef]
12. Rostami, A.; Sepidarkish, M.; Leeflang, M.M.G.; Riahi, S.M.; Nourollahpour Shiadeh, M.; Esfandyari, S.; Mokdad, A.H.; Hotez, P.J.; Gasser, R.B. SARS-CoV-2 seroprevalence worldwide: A systematic review and meta-analysis. *Clin. Microbiol. Infect.* **2020**. [CrossRef]
13. Bajema, K.L.; Wiegand, R.E.; Cuffe, K.; Patel, S.V.; Iachan, R.; Lim, T.; Lee, A.; Moyse, D.; Havers, F.P.; Harding, L.; et al. Estimated SARS-CoV-2 seroprevalence in the US as of September 2020. *JAMA Intern. Med.* **2021**, *181*, 450–460. [CrossRef] [PubMed]
14. Freeman, M.C.; Rapsinski, G.J.; Zilla, M.L.; Wheeler, S.E. Immunocompromised seroprevalence and course of illness of SARS-CoV-2 in one pediatric quaternary care center. *J. Pediatric Infect. Dis. Soc.* **2021**, *10*, 426–431. [CrossRef]
15. Wheeler, S.E.; Shurin, G.V.; Keetch, C.; Mitchell, G.; Kattel, G.; McBreen, J.; Shurin, M.R. Evaluation of SARS-CoV-2 prototype serologic test in hospitalized patients. *Clin. Biochem.* **2020**, *86*, 8–14. [CrossRef] [PubMed]
16. Azar, K.M.J.; Shen, Z.; Romanelli, R.J.; Lockhart, S.H.; Smits, K.; Robinson, S.; Brown, S.; Pressman, A.R. Disparities in outcomes among COVID-19 patients in a large health care system in California. *Health Aff. (Millwood)* **2020**, *39*, 1253–1262. [CrossRef] [PubMed]
17. Webb Hooper, M.; Napoles, A.M.; Perez-Stable, E.J. COVID-19 and racial/ethnic disparities. *JAMA* **2020**, *323*, 2466–2467. [CrossRef] [PubMed]
18. Yancy, C.W. COVID-19 and African Americans. *JAMA* **2020**, *323*, 1891–1892. [CrossRef]
19. Anaele, B.I.; Doran, C.; McIntire, R. Visualizing COVID-19 mortality rates and African-American populations in the USA and Pennsylvania. *J. Racial. Ethn. Health Disparities* **2021**. [CrossRef]
20. Grifoni, A.; Sidney, J.; Zhang, Y.; Scheuermann, R.H.; Peters, B.; Sette, A. A sequence homology and bioinformatic approach can predict candidate targets for immune responses to SARS-CoV-2. *Cell Host Microbe* **2020**, *27*, 671–680. [CrossRef]
21. Cohen, J.I.; Burbelo, P.D. Reinfection with SARS-CoV-2: Implications for vaccines. *Clin. Infect. Dis.* **2020**. [CrossRef] [PubMed]
22. Hansen, C.H.; Michlmayr, D.; Gubbels, S.M.; Mølbak, K.; Ethelberg, S. Assessment of protection against reinfection with SARS-CoV-2 among 4 million PCR-tested individuals in Denmark in 2020: A population-level observational study. *Lancet* **2021**, *397*, 1204–1212. [CrossRef]
23. Hurlburt, N.K.; Seydoux, E.; Wan, Y.H.; Edara, V.V.; Stuart, A.B.; Feng, J.; Suthar, M.S.; McGuire, A.T.; Stamatatos, L.; Pancera, M. Structural basis for potent neutralization of SARS-CoV-2 and role of antibody affinity maturation. *Nat. Commun.* **2020**, *11*, 1–7. [CrossRef] [PubMed]
24. Shi, R.; Shan, C.; Duan, X.; Chen, Z.; Liu, P.; Song, J.; Song, T.; Bi, X.; Han, C.; Wu, L.; et al. A human neutralizing antibody targets the receptor-binding site of SARS-CoV-2. *Nature* **2020**, *584*, 120–124. [CrossRef]
25. Starr, T.N.; Greaney, A.J.; Hilton, S.K.; Ellis, D.; Crawford, K.H.D.; Dingens, A.S.; Navarro, M.J.; Bowen, J.E.; Tortorici, M.A.; Walls, A.C.; et al. Deep mutational scanning of SARS-CoV-2 receptor binding domain reveals constraints on folding and ACE2 binding. *Cell* **2020**, *182*, 1295–1310. [CrossRef] [PubMed]
26. Ivanova, E.N.; Devlin, J.C.; Buus, T.B.; Koide, A.; Cornelius, A.; Samanovic, M.I.; Herrera, A.; Zhang, C.; Desvignes, L.; Odum, N.; et al. Discrete immune response signature to SARS-CoV-2 mRNA vaccination versus infection. *medRxiv* **2021**. [CrossRef]

27. Health, P.C.-VTFPDo. PA Interim Vaccine Plan V8.0. 2021. Available online: https://www.health.pa.gov/topics/Documents/Programs/Immunizations/PA%20Interim%20Vaccine%20Plan%20V.8.pdf (accessed on 6 May 2021).
28. Harmon, J.R.; Barbeau, D.J.; Nichol, S.T.; Spiropoulou, C.F.; McElroy, A.K. Rift valley fever virus vaccination induces long-lived, antigen-specific human T cell responses. *NPJ Vaccines* **2020**, *5*. [CrossRef] [PubMed]

pathogens

Article

IgG Study of Blood Sera of Patients with COVID-19

Elena Kazachinskaia [1,2], Alexander Chepurnov [1], Dmitry Shcherbakov [2], Yulia Kononova [1], Teresa Saroyan [1], Marina Gulyaeva [1,3,*], Daniil Shanshin [2], Valeriya Romanova [1], Olga Khripko [1], Michail Voevoda [1] and Alexander Shestopalov [1]

[1] Federal Research Center of Fundamental and Translational Medicine, The Federal State Budget Scientific Institution, Siberian Branch of the Russian Academy of Sciences (SB RAS), 630117 Novosibirsk, Russia; lena.kazachinskaia@mail.ru (E.K.); alexa.che.purnov@gmail.com (A.C.); yuliakononova07@yandex.ru (Y.K.); 111.st.13@rambler.ru (T.S.); valeriaromanovavr@mail.ru (V.R.); khripkoolga@gmail.com (O.K.); mvoevoda@ya.ru (M.V.); shestopalov2@mail.ru (A.S.)
[2] State Research Centre of Virology and Biotechnology "VECTOR", Federal Service for Surveillance in the Sphere, Consumers Rights Protection and Human Welfare, 630117 Novosibirsk, Russia; scherbakov_dn@vector.nsc.ru (D.S.); shanshin_dv@vector.nsc.ru (D.S.)
[3] The Department of Naturel Science, Novosibirsk State University, 630090 Novosibirsk, Russia
[*] Correspondence: mgulyaeva@gmail.com; Tel.: +7-9529136513

Abstract: The COVID-19 pandemic, which began at the end of 2019 in Wuhan, has affected 220 countries and territories to date. In the present study, we studied humoral immunity in samples of the blood sera of COVID-19 convalescents of varying severity and patients who died due to this infection, using native SARS-CoV-2 and its individual recombinant proteins. The cross-reactivity with SARS-CoV (2002) was also assessed. We used infectious and inactivated SARS-CoV-2/human/RUS/Nsk-FRCFTM-1/2020 strain, inactivated SARS-CoV strain (strain Frankfurt 1, 2002), recombinant proteins, and blood sera of patients diagnosed with COVID-19. The blood sera from patients were analyzed by the Virus Neutralization test, Immunoblotting, and ELISA. The median values and mean ± SD of titers of specific and cross-reactive antibodies in blood sera tested in ELISA were mainly distributed in the following descending order: N > trimer S > RBD. ELISA and immunoblotting revealed a high cross-activity of antibodies specific to SARS-CoV-2 with the SARS-CoV antigen (2002), mainly with the N protein. The presence of antibodies specific to RBD corresponds with the data on the neutralizing activity of blood sera. According to the neutralization test in a number of cases, higher levels of antibodies that neutralize SARS-CoV-2 were detected in blood serum taken from patients several days before their death than in convalescents with a ranging disease severity. This high level of neutralizing antibodies specific to SARS-CoV-2 in the blood sera of patients who subsequently died in hospital from COVID-19 requires a thorough study of the role of humoral immunity as well as comorbidity and other factors affecting the humoral response in this disease.

Keywords: SARS-CoV-2; COVID-19; antibody specificity; recombinant proteins; virus neutralization

Citation: Kazachinskaia, E.; Chepurnov, A.; Shcherbakov, D.; Kononova, Y.; Saroyan, T.; Gulyaeva, M.; Shanshin, D.; Romanova, V.; Khripko, O.; Voevoda, M.; et al. IgG Study of Blood Sera of Patients with COVID-19. *Pathogens* **2021**, *10*, 1421. https://doi.org/10.3390/pathogens10111421

Academic Editors: Philipp A. Ilinykh, Kai Huang and Xuguang Li

Received: 15 July 2021
Accepted: 30 October 2021
Published: 2 November 2021

1. Introduction

According to the WHO, the COVID-19 pandemic, caused by a new pathogen SARS-CoV-2 in the human population, which began at the end of 2019 in Wuhan, in China's Hubei province, has affected 220 countries and territories to date. As of 29 June 2021, the total number of reported cases worldwide exceeds 180 million, and the number of deaths worldwide is almost 4 million [1]. In many countries, quarantines and other milder strategies for preventing the spread of the infection, such as physical distancing and the use of masks, may have prevented most people from being infected. According to the opinion of some researchers, this is a paradox, as such measures leave people without immunity, susceptible to new waves of infection. Healthcare workers, the elderly, and people with medical conditions, such as cardiovascular and cerebrovascular diseases, diabetes, and neoplasms, are at a particularly high risk of infection [2,3]. It is quite possible that the modern world will not return to "pre-pandemic normality" until safe and effective vaccines have been developed and a global vaccination program has been implemented [4].

The latest reports show that most COVID-19 patients develop lymphopenia as well as pneumonia, with higher plasma levels of pro-inflammatory cytokines in severe cases, suggesting that the host immune system is involved in the pathogenesis [5]. However, there is still a very limited understanding of the immune responses, especially adaptive immune responses to SARS-CoV-2 infection. Since COVID-19 is a new disease for humanity, and the nature of the protective immune responses against it is poorly understood, it is also unclear what vaccination strategies will be the safest and most immunologically successful. It is quite possible that vaccine-induced protection may differ from natural immunity due to the virus's different evading strategies [6] and/or due to acquired humoral immunopathology in the form of antibody-dependent enhancement of infection and/or antigenic imprinting [7]. It is important to understand adaptive immunity to SARS-CoV-2, not only for vaccine development, but for interpreting COVID-19 pathogenesis and the calibration of pandemic control measures.

In this regard, data from the studies of the natural immune response to COVID-19 will contribute to COVID-19's pathogenesis, the development of effective vaccines, and therapeutic strategies. It is especially important to make clear the difference in immune responses between asymptomatic, mild, and severe cases and in the early and late stages of infection. In addition, the fact that asymptomatic and mildly suffering people develop a low level of antibody-mediated protection is important for assessing herd immunity [6]. At this moment, in the territory of Russia, this topic is not highlighted fully enough. Our investigation relates to the study of the immune response in patients with a different course of COVID-19 at the very beginning of the pandemic in one of the biggest cities of Russia under restrictions (the absence of air and railway connections with neighboring countries). It may reflect the situation at a specific point and is valuable precisely in a retrospective analysis of the development of the epidemic process in comparison with both other regions of Russia and other countries. In detail, the aim of this study was to examine humoral immunity in the samples of convalescents' blood sera with COVID-19 in a range of severities, and patients who subsequently died in hospital from this disease, using native SARS-CoV-2 and its individual recombinant proteins to visualize an individual immune "picture" of antibodies, i.e., their profile for specificity to the N proteins, the S trimer, and RBD. We consider the presence of such antibodies as part of the immune response to SARS-CoV-2. Previously, it was reported that sera from some patients could inhibit SARS-CoV-2 entry in target cells, indicating the involvement of humoral immunity due to anti-S antibodies as early as three days post symptom onset. Protein N is a major immunogen, and antibodies to it are formed in some patients who have been ill [5]. The antibodies to protein M are observed less often, and we assume that this fact would also be interesting to study. The cross-activity of antibodies with the inactivated SARS-CoV antigen (2002) was also assessed.

2. Results

To study humoral immunity against SARS-CoV-2, a random sample of 54 blood sera received from 26 convalescents (1 asymptomatic case, 13 mild cases, 1 moderate case with hospitalization for pneumonia and 1 without hospitalization, 10 severe cases, and 13 severe cases with a lethal outcome in the hospital) was used. In some cases, serum was paired from one patient. The titers of specific interaction of antibodies with these antigens were found in blood sera by the author's laboratory ELISA, using inactivated whole-cell SARS-CoV-2/human/RUS/Nsk-FRCFTM-1/2020 and SARS-CoV (strain Frankfurt 1, 2002) preparations as antigens, as well as the SARS-CoV-2 recombinant proteins—the N nucleoprotein, the full S trimer (spike) and RBD, its receptor-binding domain in the S1 region. For recombinant proteins, preparation the plasmids for the expression of the full N and full S trimer (spike) and also its part in the S1 region, RBD, were constructed with further purification of obtained proteins.

2.1. The Blood Sera Study of COVID-19 Convalescents of Varying Severity with a Favorable Outcome

During our study, it turned out that in mild cases, the median values of titers of IgG specific to the inactivated SARS-CoV-2 antigen were in the range of 1:200–1:12,800, with the most common values being 1:800 and 1:3200. High titers were found in an asymptomatic patient as well as in the case of a patient who recovered after a moderate case of COVID-19 (with unilateral pneumonia)—1:6400 (Numbers 1 and 15 on the Appendix A Figure A1). Three more recovered cases with bilateral pneumonia, or those who were treated in the intensive care unit, showed the highest titers—1:12,800, in the neutralization test with the inactivated SARS-CoV-2 antigen (Appendix A Figure A1).

The cross-activity of antibodies specific to SARS-CoV-2 with the inactivated heterologous SARS-CoV (strain Frankfurt 1, 2002) antigen (the third row on the Appendix A Figure A1, marked in green) increased in accordance with the titer found on the SARS-CoV-2 homologous antigen.

The use of recombinant SARS-CoV-2 proteins in ELISA allowed us to determine the individual profile of antibodies specific to individual viral proteins. According to ELISA data in median values (Appendix A Figure A1) and the mean ± SD of titers of specific antibodies (Table S1), the antibodies in this panel of convalescents' blood sera, from mild to severe cases in specificity and quantitatively, are mainly distributed in the following descending order: N > S trimer > RBD.

Higher IgG levels with median values in the range of 1:400 to 1:3200, detected in ELISA using the recombinant RBD protein (Appendix A Figure A1), contributed to more effective protection of infected Vero cells from the development of a cytopathic effect by 50%—from 1:20 to 1:160 in mild cases (Numbers 3, 5, 6–14), and from 1:80 to 1:≥320 in cases of moderate severity (Numbers 15, 16) and seriously ill patients (Numbers 18–32). In an asymptomatic patient (Number 1, age 32), the titer of protective antibodies was 1:80 (data on neutralization) (Table S2).

In the presence of paired sera from patients with severe cases of COVID-19 (Numbers 17–28), an increase in IgG titers specific to RBD in ELISA was observed (Appendix A Figure A1).

Obviously, the presence of antibodies specific to RBD correlates with the data on the neutralizing activity of blood sera. So, for example, the 1:200 titer of IgG specific to this site of the S glycoprotein detected in ELISA either did not contribute to protection in a severe case (Numbers 17 and 23) or corresponded to the presence of neutralizing antibodies in a titer of 1:10–1:20 in the mild cases of Numbers 2 and 4 (Table S2). In patients with a severe course of the disease from whom only single serum samples were studied in the neutralization reaction (Numbers 29–32), the median titers of protective antibodies turned out to be 1:160. In the presence of paired blood sera in patients with a severe course of the disease, a tendency towards a sharp increase in the titers of protective antibodies was revealed within a few days from complete absence to 1:160 (Numbers 23, 24) or to 1:≥320 (Numbers 17, 18). In other patients with a severe course of the disease, the titers of protective antibodies also steadily increased—for example, from 1:160 to 1:≥320 (Numbers 19, 20 and 27, 28) and from 1:80 to 1:160 (Numbers 21, 22 and 25, 26) over the course of several days, respectively (Table S2).

The immunoblotting of the SARS-CoV-2/human/RUS/Nsk-FRCFTM-1/2020 antigen with the blood serum of an asymptomatic patient, as well as those with mild and moderate cases, made it possible to detect viral target proteins for specific antibodies. Moreover, it was also possible to visually determine the efficiency/intensity of the detection of these viral proteins in this enzyme immunoassay on the nitrocellulose membrane with antibodies used for testing the same dilution of blood serum—1:100. It was clearly demonstrated that the N nucleocapsid protein of SARS-CoV-2 is the main protein target for antibodies in mild cases. Numbers 1–32 include the antibodies of asymptomatic (Number 1), mild (Numbers 2–14), and moderate patients (Numbers 15 and 16), as well as patients from the intensive care unit (Numbers 1–32) (Appendix A Figure A2A).

The antibodies specific to linear epitopes of the S-protein were not detected in all blood sera. For example, the SARS-CoV-2 S-protein dimer and trimer were detected in the serum of an asymptomatic patient (Number 1) at 180 and 270 kDa, respectively. Additionally, among mild cases of the disease (Numbers 2–14), possibly due to background levels, there are no clear bands of the S protein. Antibodies were also detected in those with moderate cases of COVID-19 (Numbers 15 and 16). In severe cases of the disease (intensive care unit) (Numbers 17–32), antibodies specific to linear epitopes of the S-protein were detected in the sera (Numbers 24–32). These findings strongly demonstrate that the intensity of the immune response significantly increased in paired sera—Numbers 23–24, 25–26, and 27–28.

The SARS-CoV (2002) nucleoprotein mostly turned out to be the only target protein for cross-interaction of antibodies specific to SARS-CoV-2 in asymptomatic, mild, moderate (Numbers 1–16), and severe cases (Numbers 17–32).

In some cases, the antibodies of some blood sera detected the M protein SARS-CoV-2 (Appendix A Figure A2A). The M protein SARS-CoV (2002) was also the target of weak cross-interaction for these antibodies (Appendix A Figure A2B).

2.2. The Blood Sera Study of Patients with a Fatal Outcome of Disease

In our study, we used sera from 13 patients with severe cases of COVID-19 with a lethal outcome in the hospital. It was shown that according to the ELISA results, the median values of specific IgG titers in the sera of these patients ranged from 1:200 to 1:25,600 (Table S3).

In the case of the paired blood sera, just as in the case of convalescents, in patients who subsequently died in the hospital, an increase in antibody titers was observed when using the inactivated SARS-CoV-2 antigen. This is especially evident in the example of three blood sera from one patient with an increase in IgG titers in ELISA from 1:6400 to 1:12,800 and 1:25,600 in blood sera for Numbers 39, 40, and 41, taken at the 15th, 10th, and 7th days before death, respectively. Moreover, these blood sera contain antibodies specific to RBD in titers of 1:1600, 1:3200, and 1:6400, corresponding to the neutralizing antibodies in the range of 1:160 to $\geq$1:320 a week before death (Table S3). The neutralizing antibodies at the median values of titers $\geq$1:320 were also found in single blood sera from two patients (Numbers 33 and 35), who subsequently died a day after the blood was taken. The sufficiently high serological titers on the inactivated antigen (1:25,600 and 1:12,800) and on the recombinant RBD (1: 25,600 and 1:3200) in ELISA were also shown (Appendix A Figure A3).

The antibodies of the blood sera of 13 patients that died from severe COVID-19 in the hospital actively cross-interact with the SARS-CoV antigen (2002) in ELISA (Appendix A Figure A3). A dependence on the value of titers with a homologous antigen was also shown, i.e., a higher titer of antibodies to SARS-CoV-2, causing a higher value of cross-activity.

According to immunoblotting data (Appendix A Figure A4A), the N protein of SARS-CoV-2 is detected by antibodies of all patients in the intensive care unit (who later died in the hospital), except for the blood serum Number 36. The presence of paired blood sera taken at different times before the death of patients makes it possible to clearly demonstrate the results of ELISA (Appendix A Figure A3) on the increase in antibody titers and the presence of antibodies specific to the linear epitopes of the S protein (dimer and trimer), and, in some cases, to the M protein. In one case, antibodies specific to the linear epitopes of the S protein monomer (90 kDa) were observed (blood sera Numbers 44 and 45). The same antibodies were also cross-reactive with analogous epitopes of the S protein monomers of SARS-CoV (2002) (Appendix A Figure A4B).

As shown in Appendix A Figure A4, in many patients in the intensive care unit (who later died in the hospital), cross-reactive antibodies were specific to both the N protein and to the S protein trimer. In serum Number 36, there were antibodies that detected linear epitopes of the S protein on the SARS-CoV antigen (2002) that did not manifest themselves in the neutralization reaction with SARS-CoV- 2 and did not reveal any bands in the SARS-CoV-2 immunoblotting.

3. Discussion

It is a fact that SARS-CoV-2 continues to spread rapidly on a global scale. Because of this, a better understanding of the relationship between the immune response and the severity of infection is needed. Serological tests are considered to be one of the main tools for epidemiological surveillance. If reliable serological data are available, the number of deaths associated with COVID-19 can help to determine the infection fatality ratio (IFR) for the disease. This knowledge can help to establish the relationship between fatalities and the total number of infections [8]. In our work, we thus provide a basis for further analysis of the pathogenesis of COVID-19, especially in severe cases, and protective immunity to SARS-CoV-2. It may also be implicated in designing an effective vaccine to protect against and treat SARS-CoV-2 infection.

Serological tests are considered to be a powerful complement to nucleic acid tests, especially for COVID-19 patients with undetectable viral RNA. Most of these tests based on the method of immunochromatography or ELISA and recombinant SARS-CoV-2 proteins have different sensitivity and specificity. In this regard, scientific groups studying the immune response of patients to COVID-19 use both commercial diagnostic kits and kits of their own production [9].

We have developed an author's ELISA test system based on antigens obtained by our scientific group. These include inactivated purified concentrated native viral preparations (SARS-CoV-2/human/RUS/Nsk-FRCFTM-1/2020 and SARS-CoV strain Frankfurt 1, 2002) and the recombinant proteins N, S, and RBD. This test system was used to determine the median titers of specific and cross-reactive antibodies in the blood sera of convalescents and patients who subsequently died from COVID-19 infection. It is shown that the cross-activity of antibodies (IgG resulting from COVID-19) with the inactivated heterologous SARS-CoV antigen (2002) increased in accordance with their titer, determined based on the homologous SARS-CoV-2 antigen in ELISA. The use of recombinant SARS-CoV-2 proteins allowed us to determine the individual profile of antibodies specific to individual viral proteins. According to the ELISA data, these antibodies in this panel of blood serum convalescents from mild to severe COVID-19, as well as in subsequent patients who died in the hospital, are distributed by specificity and quantitatively (in median titer values), mainly in the following descending order: N > trimer S > RBD.

The immunoblotting allowed us to identify linear epitopes of the main SARS-CoV-2 target proteins. In the case of a mild case of COVID-19, it is mainly a nucleoprotein. In moderate and severe cases with a favorable outcome, these are the nucleoprotein and the S trimer (270 kDa) and dimer (180 kDa). In a severe clinical course and a fatal outcome, antibodies specific to both the nucleoprotein and the S trimer (270 kDa) and dimer (180 kDa) were found in patients at different intervals before death. The literature data confirm that these coronavirus proteins are the most immunogenic of the four structural viral proteins, in addition to E (envelope) and M (membrane), to which the humoral immune response of the infected body is formed [10]. When the antibodies of some sera from surviving patients of severe COVID-19 do not interact with the linear epitopes of the S protein in immunoblotting, this cannot indicate the absence of such antibodies. In fact, it is most likely that they may be specific to the significant conformational epitopes of this viral protein, which are destroyed by abrupt electrophoretic separation.

The SARS-CoV nucleoprotein (2002) in immunoblotting was the major protein for cross-reactive antibodies in asymptomatic, mild, moderate, or severe COVID-19 with a favorable outcome.

The antibodies of some blood sera detected the M protein, which also causes the production of specific antibodies in patients, as shown by Guan M. et al. while testing the blood sera by immunoblotting, both when using an inactivated viral preparation and recombinant M protein received in eukaryotic expression system [11]. In our previous report [12], we confirmed the data of Sturman L. et al. [13], who showed that in order to visualize the matrix protein of mouse coronavirus A59, the viral preparation should not be heated prior to the electrophoretic separation of proteins.

Therefore, in this work, a viral preparation was used without heating, and it was found that not all patients suffering from COVID-19 develop an immune response to this protein. The M protein of SARS-CoV (2002) also was the target of weak cross-interactions for these antibodies.

Ouyang J. et al., summarizing the literature data, reported that titers of neutralizing antibodies in the range of 1:40–1:640 are detected in patients with COVID-19 and that the titers for donor plasma should be at least 1:80–1:160. The same authors emphasized that high titers of antibodies do not prevent a severe case of COVID-19 [14]. The results of serological testing using the recombinant S and N proteins showed that the majority of patients develop stable antibody responses to these structural viral proteins between the 17th and 23rd days after the disease started. A slower but stronger antibody response has been observed in severe cases. It was noted that in patients with more severe cases of COVID-19, antibodies specific to the S protein showed high titers [15].

According to our data, it is shown that the presence of antibodies specific to RBD mostly correlates with the data on the neutralizing activity of blood sera. With almost all deaths, barring a few exceptions, the titers of neutralizing antibodies were at the level of severely ill convalescents and even higher, but this did not save this group of patients. Most likely, this happened, as described in the literature, due to multiple organ disorders in the infected body as a result of SARS-CoV-2's immunopathogenesis. This was expressed in the fact that viral activation of dendritic cells and macrophages in lymphoid tissues leads to an excessive uncontrolled anti-inflammatory response, the so-called cytokine storm [16]. In addition, data have already been obtained on the dependence of the severity of COVID-19, which was established at the beginning of the pandemic, with increased IgG titers when compared with a low level of antibodies of this class in convalescents. The mechanism responsible for the immunopathological IgG response remains unclear [17].

It is not known whether patients from our random sample had previous contact with other seasonal coronaviruses (CoVs), for which antibodies are widely present in the human population. Several conserved regions have already been identified in the S2 domains of four known CoVs, namely 229E, NL63, OC43, HKUI, and SARS-CoV-2 [18], which is believed to be the cause of the antibody-dependent increase in the infection [19]. However, under ELISA and immunoblotting, when using the N recombinant protein of SARS-CoV-2 obtained in *E. coli*, cross-activity with this protein was observed only for antibodies of blood sera of patients who had previously recovered from SARS-CoV, but not with other CoVs [20]. In addition, the authors analyzed the amino acid sequence (aa) of the N protein of known CoVs isolated from humans. The N protein of SARS-CoV-2 turned out to have a homology of 19.1, 20.0, 26.5, 27.6, 46.1, and 90.5% with 229E, NL63, OC43, HKUI, MERS-CoV, and SARS-CoV, respectively [20].

In this study, we did not take into account the comorbidities of the patients. This aspect is also important for understanding viral pathogenesis. It is known that risk factors for a fatal outcome in COVID-19, in addition to male sex and old age [12], include diabetes mellitus and hyperglycemia that develops when infected with SARS-CoV-2 [21], cardio-vascular diseases [22], etc. The analysis of the neutralizing activity of specific antibodies is an important stage in the study of the immune response of patients with COVID-19. Therefore, an extensive study of the immune response to SARS-CoV-2 in the population must be associated with such data for each patient with COVID-19, and this shows a further necessity for deeper investigations into the role of humoral immunity in COVID-19 disease.

A varied course of coronavirus infections was observed over the course of a year and a half of fighting the pandemic. In this study, we highlighted the immune response in patients with a different course of COVID-19 at the very beginning of the pandemic in one of the major cities of Russia under restrictions (the absence of air and railway connections with neighboring countries). It may reflect the situation at a specific point and is valuable precisely in a retrospective analysis of the development of the epidemic process. We tried to get closer to understanding the role of humoral immunity in the development of pathogenesis, which still requires careful study.

4. Materials and Methods

4.1. Viral Preparations

SARS-CoV-2/human/RUS/Nsk-FRCFTM-1/2020 strain was isolated from a clinical sample of a nasopharyngeal swab from the patient in Novosibirsk. The virus was purified and concentrated, as described [12]. Both SARS-CoV-2/human/RUS/Nsk-FRCFTM-1/2020 and an antigen of the SARS-CoV strain Frankfurt 1 (2002) [23] were stored at the Federal Research Center of Fundamental and Translational Medicine SB RAS at $-80\,^{\circ}$C. Viral preparations with an infectious titer of 6 lg tissue cytopathic doses (TCPD$_{50}$/$_{mL}$) were inactivated (in a 1:1 ratio) in a lysis buffer used for electrophoresis (composition for 2 mL: 1 M Tris-HCl with pH 6.8–0.5 mL; 10% sodium dodecyl sulfate (SDS) solution—0.8 mL; glycerin—0.2 mL; mercaptoethanol—0.2 mL; 0.4% Bromphenol Blue Na-sulf—0.1 mL).

4.2. Recombinant Proteins

On the basis of the nucleotide sequences of SARS-CoV-2 presented in GenBank: MN908947 [24], the plasmids for the expression of the full N and full S trimer (spike) and also its part in the S1 region, RBD, were constructed. Recombinant proteins of the S trimer and RBD were obtained in eukaryotic cells lines of female hamster ovarian (CHO-K1), and the nucleoproteins were obtained in *E. coli*. All recombinant proteins were purified with affinity chromatography.

4.3. Blood Sera

In this study, we utilized: 1. the convalescents' blood sera from the employees of the Federal Research Center of the Fundamental and Translational Medicine SB RAS and their relatives who had suffered from mild cases of COVID-19, including asymptomatic cases, as well as moderate COVID-19 cases with pneumonia; 2. blood sera from intensive care unit patients with severe COVID-19, which were retrieved from the Infectious Diseases Clinical Hospital No. 1 in Novosibirsk. The diagnosis of COVID-19 in patients was based on the presence of SARS-CoV-2 RNA, which was detected using the RealBest SARS-CoV-2 RNA kit (Vector-Best, Novosibirsk, Russia) in a certified laboratory for the diagnosis of COVID-19 at the Federal Research Center of the Fundamental and Translational Medicine SB RAS. Permission to use the clinical material was obtained from the Ethics Committee of the Research Center (Protocol No. 17 as of 17 June 2020). The blood sera of convalescents and patients from the Infectious Disease Hospital were tested for the absence of SARS-CoV-2 RNA before conducting this study.

4.4. ELISA

IgG specific to SARS-CoV-2 was found using the author's laboratory test system. Polystyrene plates (Nunk) were used for ELISA. The wells of plates were sensitized with both SARS-CoV-2 and SARS-CoV (strain Frankfurt 1, 2002) viral proteins and recombinant antigens at a concentration of 2 µg/mL in 100 µL of 0.05 M sodium phosphate buffer solution (pH 8.0) at 220 $^{\circ}$C for 18 h. Nonspecific binding sites were saturated with 1% casein solution (Sigma, St. Louis, MO, USA) in TBST (Tris Buffered Saline with Tween containing 0.15 M NaCl; 0.02 M Tris-HCl pH 7.4; 0.05% Tween-20) within 45 min at 37 $^{\circ}$C. Then, the antigens were incubated with blood sera (with their preliminary depletion of the nonspecific background by 5% of the volume of cell lysates, on which culture the corresponding antigens were produced) with a dilution of 1/100 (double step for titration) in a 0.5% casein solution for 1 h at 37 $^{\circ}$C. The peroxidase conjugate of anti-species antibodies (Gout anti-human IgG, Sigma) was used at a working dilution of 1/6000 in 0.5% casein solution with incubation for 1 h at 37 $^{\circ}$C. The immune response was manifested using a liquid substrate based on TMB (3,3′, 5,5′-tetramethylbenzidine). The reaction was blocked by adding 100 µL of 1 N HCl to each well. The absorbency of the substrate–indicator mixture was measured on a Uniscan spectrophotometer at a wavelength of 450 nm. A lysate of uninfected Vero cells and the blood serum of healthy people were used as a negative control of the antigen. The results of ELISA detected in three repetitions in two independent experiments were calculated using median values.

4.5. Immunoblotting

The SARS-CoV-2 and SARS-CoV (strain Frankfurt 1, 2002) viral proteins were separated in one wide "pocket" by electrophoresis in a 10% polyacrylamide gel (PAGE) supplemented with sodium dodecyl sulfate (SDS) and transferred to a nitrocellulose membrane (Millipore) in equipment (Cole-Parmer, Vernon Hills, IL, USA) for incubating blotting membranes for 5 h at 50 V, in a 0.025 M Tris-HCl buffer containing 0.192 M glycine (pH 8.3) and 20% ethanol. The sites of nonspecific binding were saturated with a 1% casein solution for 1.5 h at 20–22 °C. The whole membrane containing viral proteins was cut into separate strips, then numbered and incubated in separate containers for 4 h at 20–22 °C, with human blood sera (dilution 1:100) in a TSBT buffer containing 0.5% casein. After being washed in the TSBT buffer, the membrane strips were treated with anti-species antibodies labeled with horseradish peroxidase (Gout anti-human IgG, Sigma) at a working dilution of 1/6000 in 0.5% casein solution for 2 h at 37 °C. Then, the membrane strips were washed with TSBT buffer and developed in a chromagen solution (1 mg/mL 3.3 diaminobenzidine tetrahydrochloride in 50 mM Tris-HCl buffer (pH 7.4) containing 0.145 M NaCl, 20% ethanol, and 0.03% hydrogen peroxide). The reaction was stopped by washing the membrane strips in TSBT buffer. The specific interaction of the antibodies with viral proteins was manifested by the brown color of the stripes.

4.6. Virus Neutralization

Virus neutralization with antibodies was carried out in accordance with the generally accepted method, as described [25]. The titer of the infectious virus was expressed in $TCPD_{50}/mL$ (tissue cytopathic dose of the virus causing a 50% cytopathic effect on the infected cells). It was found by titrating the viral preparation on the Vero cells monolayer (African green monkey kidney cell culture) grown in 96-well culture plates (Corning, Glendale, AZ, USA). Before use, blood serum was inactivated by heating at 56 °C for 30 min to inactivate the antiviral effect of complement proteins. Before applying it to a monolayer of cells, blood serum ranging from 1 to 20 with two dilutions was preincubated with the infectious SARS-CoV-2/human/RUS/Nsk-FRCFTM-1/2020 strain in a titer of 103 $TCPD_{50}/mL$ (50% tissue cytopathic doses in mL) for 1 h at 37 °C in a nutrient medium containing a 2% blood serum (heated at 56 °C for 30 min) of cattle. Then, it was applied to a monolayer of cell culture in three replicates. After the incubation of the mixture of antibodies with the virus for 1 h at 37 °C, the monolayer of cells was washed and left in a nutrient medium containing 2% cattle blood sera until a cytopathogenic effect was observed in control wells containing infected cells. To observe the infected and control cells in dynamics, an inverted microscope Mikromed I (Mikromed, Sankt Petersburg, Russia) was used at $10\times$ magnification. The cells were fixed for 30 minutes with a formaldehyde solution and a 0.05% crystalline violet solution with 20% alcohol, as described [26]. Then, the liquid from the wells was removed and washed with water. The results were recorded visually. The neutralizing antibodies titer was considered the final dilution of blood serum at which cells were protected from the cytopathogenic effect in 50% of the wells. The neutralizing antibodies titer, which was detected in three repetitions in two independent experiments, was calculated using the median results.

Supplementary Materials: The following are available online at https://www.mdpi.com/article/10.3390/pathogens10111421/s1, Table S1: Specific interaction of inactivated viruses and recombinant proteins in IgG ELISA of convalescents and patients died from COVID-19; Table S2: The neutralizing activity of the blood serum of convalescents and patients with COVID-19; Table S3. The neutralizing activity of the blood serum of patients who subsequently died in hospital from COVID-19.

Author Contributions: Conceptualization, A.S.; methodology, E.K., D.S. (Dmitry Shcherbakov) and A.C.; validation, A.C., A.S. and E.K.; formal analysis, O.K. and D.S. (Daniil Shanshin); investigation,

A.C., T.S., Y.K. and V.R.; resources, M.V.; data curation, E.K.; writing—original draft preparation, E.K.; writing—review and editing, M.G. and E.K.; supervision, M.G. and M.V.; project administration, A.S. All authors have read and agreed to the published version of the manuscript.

Funding: The reported study was funded by RFBR according to the research project No 20-54-80012.

Institutional Review Board Statement: The study was conducted according to the guidelines of the Declaration of Helsinki and approved by the Ethics Committee of the Federal Research Center for Fundamental and Translational Medicine (Protocol No. 17 as of 17 June 2020) and from the Infectious Disease Hospital #1 (Protocol No. S/025 as of 12 May 2020).

Informed Consent Statement: Informed consent was obtained from all subjects involved in the study.

Data Availability Statement: The data presented in this study are available in the main text, figures, tables and supplementary material.

Acknowledgments: We gratefully acknowledge FBHF City Infectious Diseases Clinical Hospital No. 1 (Novosibirsk) and its staff for providing sera from patients used in experimental analysis.

Conflicts of Interest: The authors declare no conflict of interest.

Appendix A

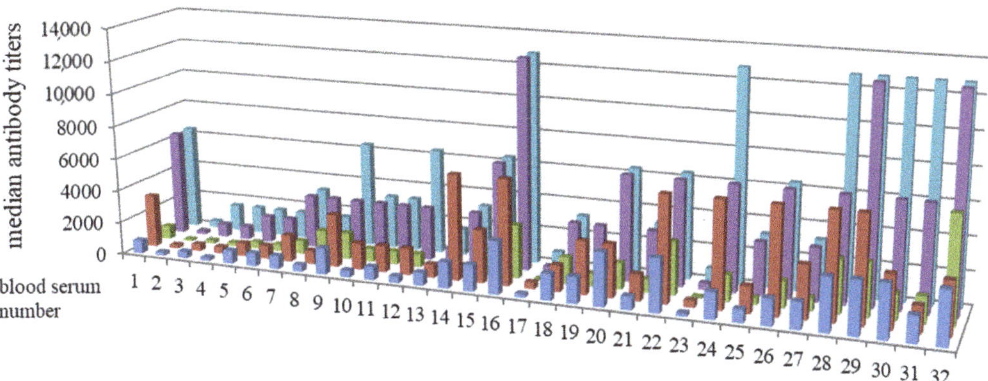

Figure A1. Specific interaction of IgG convalescents with inactivated viral preparations and recombinant proteins in ELISA. Note. Blood sera: No. 1—an asymptomatic case; No. 2–14—mild cases, No. 15—moderate case with unilateral pneumonia; No. 16—moderate case with hospitalization for bilateral pneumonia; No.17–32—severe cases of patients who were treated in the intensive care unit. Data on the neutralizing activity of these blood sera are presented in Tables S1 and S2.

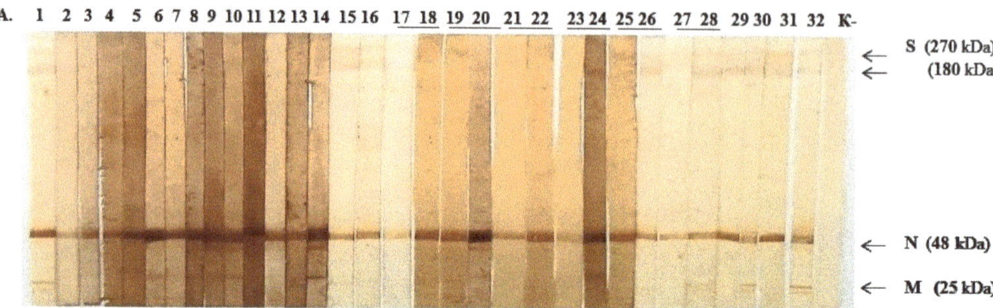

Figure A2. Immunoblotting of viral proteins of the SARS-CoV-2/human/RUS/Nsk-FRCFTM-1/2020 (2020) and SARS-CoV (2002) strains with antibodies of convalescents' blood sera (asymptomatic, mild or moderate cases of COVID-19, as well as patients in the intensive care unit). (**A**). The SARS-CoV-2/human/RUS/Nsk-FRCFTM-1/2020 antigen on the membrane. (**B**). The SARS-CoV antigen (2002) on the membrane. Blood sera: No. 1—an asymptomatic case; No. 2–14—mild cases, No. 15—a moderate case with unilateral pneumonia; No. 16—a moderate case with hospitalization for bilateral pneumonia; No. 17–32—severe cases of patients in the intensive care unit. K- (negative control)—blood serum of a healthy person. The numbers of paired blood sera are underlined.

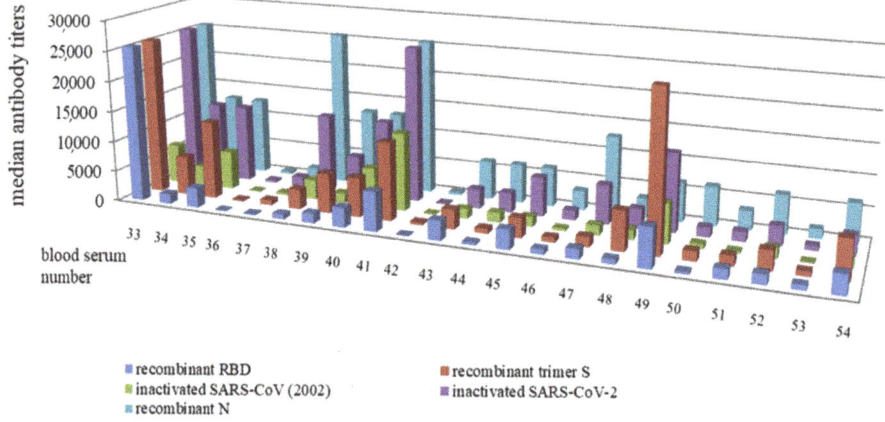

Figure A3. Titers of specific IgG interactions in patients with severe cases of COVID-19 who subsequently died in hospital, with inactivated viral preparations and recombinant proteins in ELISA. Note: data on blood sera numbers are presented in Tables S1 and S3.

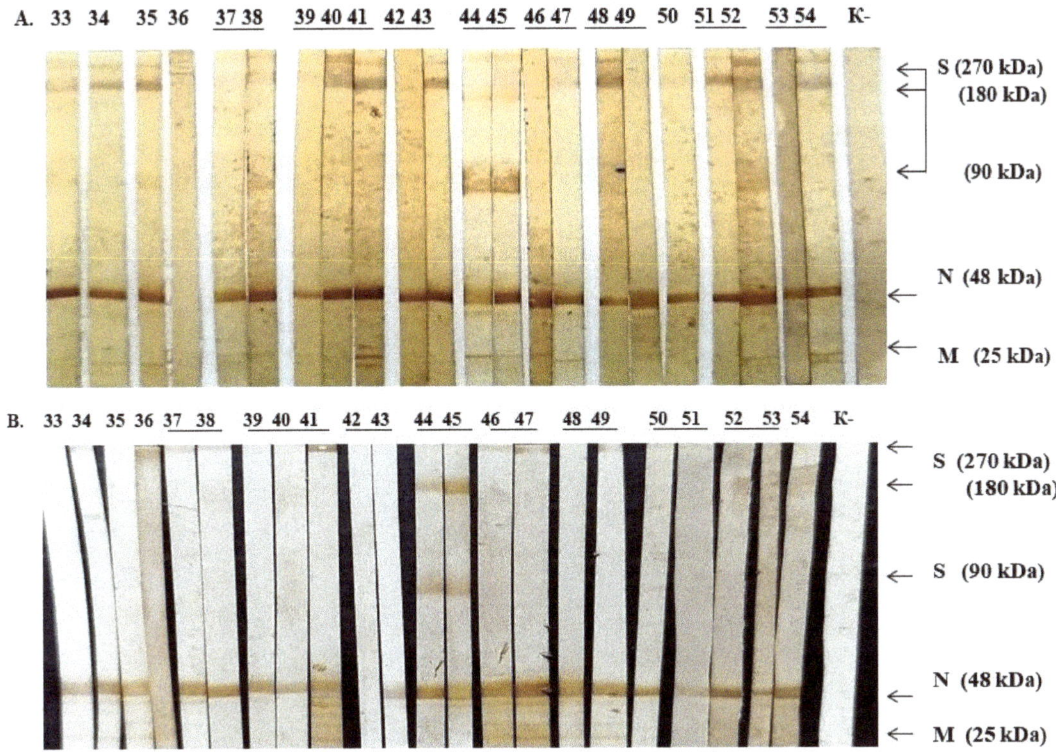

Figure A4. Immunoblotting of viral proteins of the SARS-CoV-2/human/RUS/Nsk-FRCFTM-1/2020 (2020) and SARS-CoV (2002) strains with antibodies of blood sera of patients who were treated in the intensive care unit diagnosed with COVID-19 (later died in the hospital). (**A**) SARS-CoV-2/human/RUS/Nsk-FRCFTM-1/2020 antigen on the membrane. (**B**) SARS-CoV antigen (2002) on the membrane. *K-* (negative control)—blood serum of a healthy person. The numbers of paired blood sera are underlined.

References

1. WHO. Report World Health Organization. WHO Coronavirus Disease (COVID-19) Dashboard. Available online: https://www.who.int/emergencies/diseases/novel-coronavirus-2019 (accessed on 4 July 2021).
2. Chou, R.; Dana, T.; Buckley, D.I.; Selph, S.; Fu, R.; Totten, A.M. Epidemiology of and Risk Factors for Coronavirus Infection in Health Care Workers. *Ann. Intern. Med.* **2020**, *173*, 120–136. [CrossRef]
3. Sanche, S.; Lin, Y.T.; Xu, C.; Romero-Severson, E.; Hengartner, N.; Ke, R. High Contagiousness and Rapid Spread of Severe Acute Respiratory Syndrome Coronavirus 2. *Emerg. Infect. Dis.* **2020**, *26*, 1470–1477. [CrossRef]
4. WHO. Draft Landscape and Tracker of COVID-19 Candidate Vaccines. Available online: https://www.who.int/publications/m/item/draft-landscape-of-covid-19-candidate-vaccines (accessed on 25 January 2021).
5. Ni, L.; Ye, F.; Cheng, M.-L.; Feng, Y.; Deng, Y.-Q.; Zhao, H.; Wei, P.; Ge, J.; Gou, M.; Li, X.; et al. Detection of SARS-CoV-2-Specific Humoral and Cellular Immunity in COVID-19 Convalescent Individuals. *Immunity* **2020**, *52*, 971–977.e3. [CrossRef]
6. Jeyanathan, M.; Afkhami, S.; Smaill, F.; Miller, M.; Lichty, B.D.; Xing, Z. Immunological considerations for COVID-19 vaccine strategies. *Nat. Rev. Immunol.* **2020**, *20*, 1–18. [CrossRef]
7. Супотницкий, M.B. COVID-19 Pandemic as an Indicator of «Blank Spots» in Epidemiology and Infectious Pathology. *J. NBC Prot. Corps* **2020**, *4*, 338–373. [CrossRef]
8. O'Driscoll, M.; Dos Santos, G.R.; Wang, L.; Cummings, D.A.T.; Azman, A.S.; Paireau, J.; Fontanet, A.; Cauchemez, S.; Salje, H. Age-specific mortality and immunity patterns of SARS-CoV-2. *Nature* **2020**, *590*, 140–145. [CrossRef]
9. Huang, C.; Wang, Y.; Li, X.; Ren, L.; Zhao, J.; Hu, Y.; Zhang, L.; Fan, G.; Xu, J.; Gu, X.; et al. Clinical features of patients infected with 2019 novel coronavirus in Wuhan, China. *Lancet* **2020**, *395*, 497–506. [CrossRef]
10. Qiu, M.; Shi, Y.; Guo, Z.; Chen, Z.; He, J.; Chen, R.; Zhou, D.; Dai, E.; Wang, X.; Si, B.; et al. Antibody responses to individual proteins of SARS coronavirus and their neutralization activities. *Microbes Infect.* **2005**, *7*, 882–889. [CrossRef]

11. Guan, M.; Chen, H.Y.; Tan, P.H.; Shen, S.; Goh, P.-Y.; Tan, Y.-J.; Pang, P.H.; Lu, Y.; Fong, P.Y.; Chin, D. Use of Viral Lysate Antigen Combined with Recombinant Protein in Western Immunoblot Assay as Confirmatory Test for Serodiagnosis of Severe Acute Respiratory Syndrome. *Clin. Vaccine Immunol.* **2004**, *11*, 1148–1153. [CrossRef]

12. Chepurnov, A.A.; Sharshov, K.A.; Kazachinskaya, E.I.; Kononova, Y.V.; Kazachkova, E.A.; Khripko, O.P.; Yurchenko, K.S.; Alekseev, A.Y.; Voevoda, M.I.; Shestopalov, A.M. Antigenic properties of sARs-CoV-2/human/RUs/nsk-FRCFtM-1/2020 coronavirus isolate from a patient in novosibirsk. *J. Infectology* **2020**, *12*, 42–50. [CrossRef]

13. Sturman, L.S. Characterization of a coronavirus. *Virology* **1977**, *77*, 637–649. [CrossRef]

14. Ouyang, J.; Isnard, S.; Lin, J.; Fombuena, B.; Peng, X.; Routy, J.-P.; Chen, Y. Convalescent Plasma: The Relay Baton in the Race for Coronavirus Disease 2019 Treatment. *Front. Immunol.* **2020**, *11*, 2424. [CrossRef]

15. Qu, J.; Wu, C.; Li, X.; Zhang, G.; Jiang, Z.; Li, X.; Zhu, Q.; Liu, L. Profile of Immunoglobulin G and IgM Antibodies Against Severe Acute Respiratory Syndrome Coronavirus 2 (SARS-CoV-2). *Clin. Infect. Dis.* **2020**, *71*, 2255–2258. [CrossRef]

16. Zhou, G.; Zhao, Q. Perspectives on therapeutic neutralizing antibodies against the Novel Coronavirus SARS-CoV-2. *Int. J. Biol. Sci.* **2020**, *16*, 1718–1723. [CrossRef]

17. Zhang, B.; Zhou, X.; Zhu, C.; Song, Y.; Feng, F.; Qiu, Y.; Feng, J.; Jia, Q.; Song, Q.; Zhu, B.; et al. Immune Phenotyping Based on the Neutrophil-to-Lymphocyte Ratio and IgG Level Predicts Disease Severity and Outcome for Patients With COVID-19. *Front. Mol. Biosci.* **2020**, *7*, 157. [CrossRef]

18. Lan, J.; Ge, J.; Yu, J.; Shan, S.; Zhou, H.; Fan, S.; Zhang, Q.; Shi, X.; Wang, Q.; Zhang, L.; et al. Structure of the SARS-CoV-2 spike receptor-binding domain bound to the ACE2 receptor. *Nature* **2020**, *581*, 215–220. [CrossRef]

19. Tetro, J.A. Is COVID-19 receiving ADE from other coronaviruses? *Microbes Infect.* **2020**, *22*, 72–73. [CrossRef]

20. Guo, L.; Ren, L.; Yang, S.; Xiao, M.; Chang, D.; Yang, F.; Cruz, C.S.D.; Wang, Y.; Wu, C.; Xiao, Y.; et al. Profiling Early Humoral Response to Diagnose Novel Coronavirus Disease (COVID-19). *Clin. Infect. Dis.* **2020**, *71*, 778–785. [CrossRef]

21. Lim, S.; Bae, J.H.; Kwon, H.-S.; Nauck, M.A. COVID-19 and diabetes mellitus: From pathophysiology to clinical management. *Nat. Rev. Endocrinol.* **2020**, *17*, 11–30. [CrossRef]

22. Azevedo, R.B.; Botelho, B.G.; De Hollanda, J.V.G.; Ferreira, L.V.L.; De Andrade, L.Z.J.; Oei, S.S.M.L.; Mello, T.D.S.; Muxfeldt, E.S. Covid-19 and the cardiovascular system: A comprehensive review. *J. Hum. Hypertens.* **2020**, *35*, 4–11. [CrossRef]

23. Agafonov, A.P.; Gus'Kov, A.A.; Ternovoi, V.A.; Ryabchikova, E.I.; Durymanov, A.G.; Vinogradov, I.V.; Maksimov, N.L.; Ignat'Ev, G.M.; Nechaeva, E.A.; Netesov, C.M.O.R.S.V. Primary Characterization of SARS Coronavirus Strain Frankfurt 1. *Dokl. Biol. Sci.* **2004**, *394*, 58–60. [CrossRef]

24. Wu, F.; Zhao, S.; Yu, B.; Chen, Y.-M.; Wang, W.; Song, Z.-G.; Hu, Y.; Tao, Z.-W.; Tian, J.-H.; Pei, Y.-Y.; et al. A new coronavirus associated with human respiratory disease in China. *Nature* **2020**, *579*, 265–269. [CrossRef]

25. Van Doremalen, N.; Lambe, T.; Spencer, A.; Belij-Rammerstorfer, S.; Purushotham, J.N.; Port, J.R.; Avanzato, V.A.; Bushmaker, T.; Flaxman, A.; Ulaszewska, M.; et al. ChAdOx1 nCoV-19 vaccine prevents SARS-CoV-2 pneumonia in rhesus macaques. *Nature* **2020**, *586*, 1–8. [CrossRef]

26. Case, J.B.; Bailey, A.L.; Kim, A.S.; Chen, R.E.; Diamond, M.S. Growth, detection, quantification, and inactivation of SARS-CoV-2. *Virology* **2020**, *548*, 39–48. [CrossRef]

Relative COVID-19 Viral Persistence and Antibody Kinetics

Chung-Guei Huang [1,2,3,†], Avijit Dutta [3,4,†], Ching-Tai Huang [4,5], Pi-Yueh Chang [1,2], Mei-Jen Hsiao [1,2], Yu-Chia Hsieh [6,7], Shu-Min Lin [8,9], Shin-Ru Shih [1,2,3], Kuo-Chien Tsao [1,2] and Cheng-Ta Yang [8,9,*]

[1] Department of Laboratory Medicine, Chang Gung Memorial Hospital, Taoyuan 33333, Taiwan; joyce@cgmh.org.tw (C.-G.H.); changpy@cgmh.org.tw (P.-Y.C.); m9205025@stmail.cgu.edu.tw (M.-J.H.); srshih@mail.cgu.edu.tw (S.-R.S.); kctsao@adm.cgmh.org.tw (K.-C.T.)
[2] Department of Medical Biotechnology and Laboratory Science, College of Medicine, Chang Gung University, Taoyuan 33302, Taiwan
[3] Research Center for Emerging Viral Infections, College of Medicine, Chang Gung University, Taoyuan 33302, Taiwan; duttijiva@gmail.com
[4] Division of Infectious Diseases, Department of Medicine, Chang Gung Memorial Hospital, Taoyuan 33333, Taiwan; chingtaihuang@gmail.com
[5] Division of Infectious Diseases, Department of Medicine, College of Medicine, Chang Gung University, Taoyuan 33302, Taiwan
[6] Division of Infectious Diseases, Department of Pediatrics, Chang Gung Memorial Hospital, Taoyuan 33333, Taiwan; yuchiahsieh@gmail.com
[7] Division of Infectious Diseases, Department of Pediatrics, College of Medicine, Chang Gung University, Taoyuan 33302, Taiwan
[8] Department of Thoracic Medicine, Chang Gung Memorial Hospital, Taoyuan 33333, Taiwan; smlin100@gmail.com
[9] Department of Respiratory Therapy, College of Medicine, Chang Gung University, Taoyuan 33302, Taiwan
* Correspondence: yang1946@cgmh.org.tw; Tel.: +886-3-3281200
† These authors contributed equally to this work.

Citation: Huang, C.-G.; Dutta, A.; Huang, C.-T.; Chang, P.-Y.; Hsiao, M.-J.; Hsieh, Y.-C.; Lin, S.-M.; Shih, S.-R.; Tsao, K.-C.; Yang, C.-T. Relative COVID-19 Viral Persistence and Antibody Kinetics. *Pathogens* **2021**, *10*, 752. https://doi.org/10.3390/pathogens10060752

Academic Editors: Philipp A. Ilinykh and Kai Huang

Received: 10 May 2021
Accepted: 10 June 2021
Published: 13 June 2021

Publisher's Note: MDPI stays neutral with regard to jurisdictional claims in published maps and institutional affiliations.

Abstract: A total of 15 RT-PCR confirmed COVID-19 patients were admitted to our hospital during the in-itial outbreak in Taiwan. The average time of virus clearance was delayed in seven patients, 24.14 ± 4.33 days compared to 10.25 ± 0.56 days post-symptom onset (PSO) in the other eight pa-tients. There was strong antibody response in patients with viral persistence at the pharynx, with peak values of serum antibody 677.2 ± 217.8 vs. 76.70 ± 32.11 in patients with delayed versus rapid virus clearance. The patients with delayed viral clearance had excessive antibodies of compromised quality in an early stage with the delay in peak virus neutralization efficacy, 34.14 ± 7.15 versus 12.50 ± 2.35 days PSO in patients with rapid virus clearance. Weak antibody re-sponse of patients with rapid viral clearance was also effective, with substantial and comparable neutralization efficacy, 35.70 ± 8.78 versus 41.37 ± 11.49 of patients with delayed virus clearance. Human Cytokine 48-Plex Screening of the serial sera samples revealed elevated concentrations of proinflammatory cytokines and chemokines in a deceased patient with delayed virus clear-ance and severe disease. The levels were comparatively less in the other two patients who suf-fered from severe disease but eventually survived.

Keywords: COVID-19; viral persistence; serum antibody; neutralization efficacy; cytokine profile

1. Introduction

In late January 2020, we started to treat real-time polymerase chain reaction (RT-PCR)-confirmed COVID-19 patients. We are a medical center that typically cares for patients with moderate to severe diseases. Because of the low prevalence of COVID-19 in Taiwan and local government policy, however, we also admitted COVID-19 patients with mild disease or even those without symptoms for inpatient care. Serial RT-PCR tracking of pharyngeal samples was performed throughout each patient's hospital course. With informed consent from patients or their families, our research was conducted using serum samples remaining after routine medical tests. We used two ELISA-based kits [1–3] to detect anti-spike protein IgG antibodies, and the results of the two were concordant. The tests were semi-quantitative and measured antibody concentrations relative to a cut-off point value in serial dilutions of serum samples. We cultured virus strains from our patient samples and used one of the strains to quantify the neutralization valence of serum samples in our Biosafety Level-3 laboratory. We also measured cytokines and chemokines in the serial sera samples of four patients using the Bio-Plex Pro Human Cytokine 48-Plex Screening kit from Bio-Rad. By mid-March 2020, we had collected serum samples from 15 consecutive patients. As this was a single-center study, we also collected a complete medical record including detailed travel, occupation, contacts, and cluster history.

2. Results

2.1. Relative Viral Persistence

A total of 15 RT-PCR confirmed COVID-19 patients were admitted to the Linkou campus of Chang Gung Memorial Hospital during the study period as a result of the initial outbreak in Taiwan. All 15 COVID-19 patients were included in this study (Table 1). They suffered from a range of asymptomatic to severe diseases, and virus clearance varied from day 7 to day 49 in these patients (Figure 1). We divided our 15 patients into 2 groups according to whether they may clear the virus within 2 weeks post-symptom onset (PSO). Of the participants involved, 7 patients cleared the virus after two weeks, and 8 patients eradicated the virus within two weeks. For the 7 patients with delayed clearance, the average time of virus clearance was 24.14 ± 4.33 days PSO, and for the 8 patients with rapid clearance, the average time to clear the virus was 10.25 ± 0.56 days PSO ($p = 0.0046$; Figure 2A). There was a significant difference in the age of the two groups of patients. Older patients could not eradicate the virus in time (60.14 ± 3.58 vs. 38.25 ± 5.28 years, $p = 0.0054$). The slope of the virus decline was flat in patients with delayed clearance, in contrast to a sharp slope of virus decline in patients with rapid clearance. We then used the area under curve (AUC) analysis to compare the effective existence of the virus. There was 52% more area under the curve for patients with delayed clearance compared to patients with rapid clearance. The Ct values of E gene at disease presentation were comparable between both the groups of patients (26.02 ± 3.15, $n = 7$ vs. 25.17 ± 4.57, $n = 8$, $p = 0.693$; Figure 2A). All these differences between the two groups were not associated with the initial virus burden at disease presentation.

Table 1. Demographic and clinical characteristics of the patients with COVID-19.

Patient Cohort	With Virus Persistence (Mean $\pm$ SD)	With Rapid Virus Clearance (Mean $\pm$ SD)
Number	7 (2 Male/5 Female)	8 (3 Male/5 Female)
Age (Years)	60.14 ± 3.58	38.25 ± 5.28
ICU assistance	3/7	1/8
ECMO support	2/7	1/8
Mortality	1/7 (14.3%)	1/8 (12.5%)
Virus burden at presentation:		
E gene, Ct value	26.02 ± 3.59	25.75 ± 2.67
RdRp1 gene, Ct value	27.87 ± 5.68	25.88 ± 4.53
RdRp2 gene, Ct value	26.35 ± 5.27	25.61 ± 4.73
N gene, Ct value	28.52 ± 5.34	27.99 ± 4.04
Virus clearance		
Days, PSO	24.14 ± 4.33	10.25 ± 0.56

Table 1. *Cont.*

Patient Cohort	With Virus Persistence (Mean ± SD)	With Rapid Virus Clearance (Mean ± SD)
Anti-spike IgG antibody Response in sera:		
Peak levels	677.2 ± 217.8	76.70 ± 32.11
Time for peak (Days, PSO)	17.43 ± 2.61	11.13 ± 2.48
Virus-neutralizing antibodies		
Capacity/unit sera (Peak levels)	168.00 ± 63.42	29.68 ± 14.82
Time for peak (Days, PSO)	30.43 ± 7.80	12.20 ± 5.17
Efficacy/unit antibody (Peak levels)	41.37 ± 11.49	35.70 ± 8.78
Time for peak (Days, PSO)	34.14 ± 7.15	12.50 ± 2.35

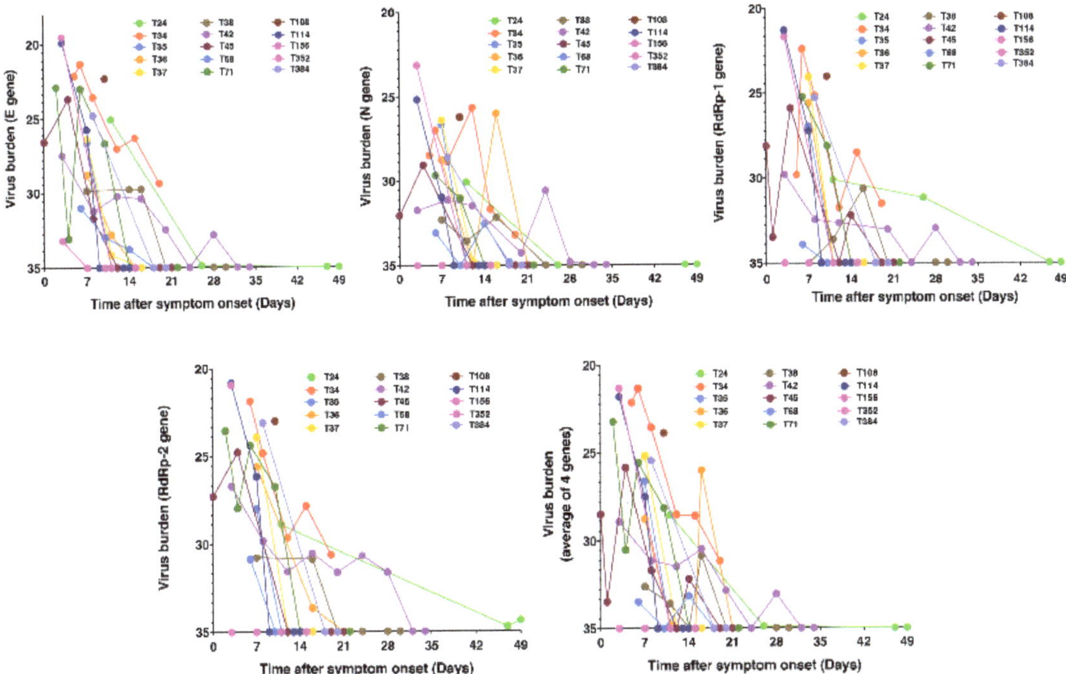

Figure 1. *Pharyngeal virus load and clearance.* Kinetics of virus burden in terms of E, N, RdRp-1, and RdRp-2 gene Ct values at stated time points after symptom onset. Pharyngeal samples were collected and RT-PCR tests were performed as described in the text. (The patients are tagged by our national serial number of COVID-19 cases).

2.2. Antibody Kinetics and Relative Viral Persistence

The serum antibody levels were much higher in patients with delayed virus clearance. The peak values of serum antibodies were 677.2 ± 217.8 vs. 76.70 ± 32.11 in patients with delayed versus rapid virus clearance ($p < 0.0001$; Figure 2B). With comparable levels of initial antibody response, the peak antibody response also emerged later in the patients with delayed virus clearance. The time from symptom onset to the time of peak serum antibody levels was 17.43 ± 2.61 days PSO and 11.13 ± 2.48 days PSO in delayed and rapid clearance groups, respectively ($p = 0.0004$; Figure 2B).

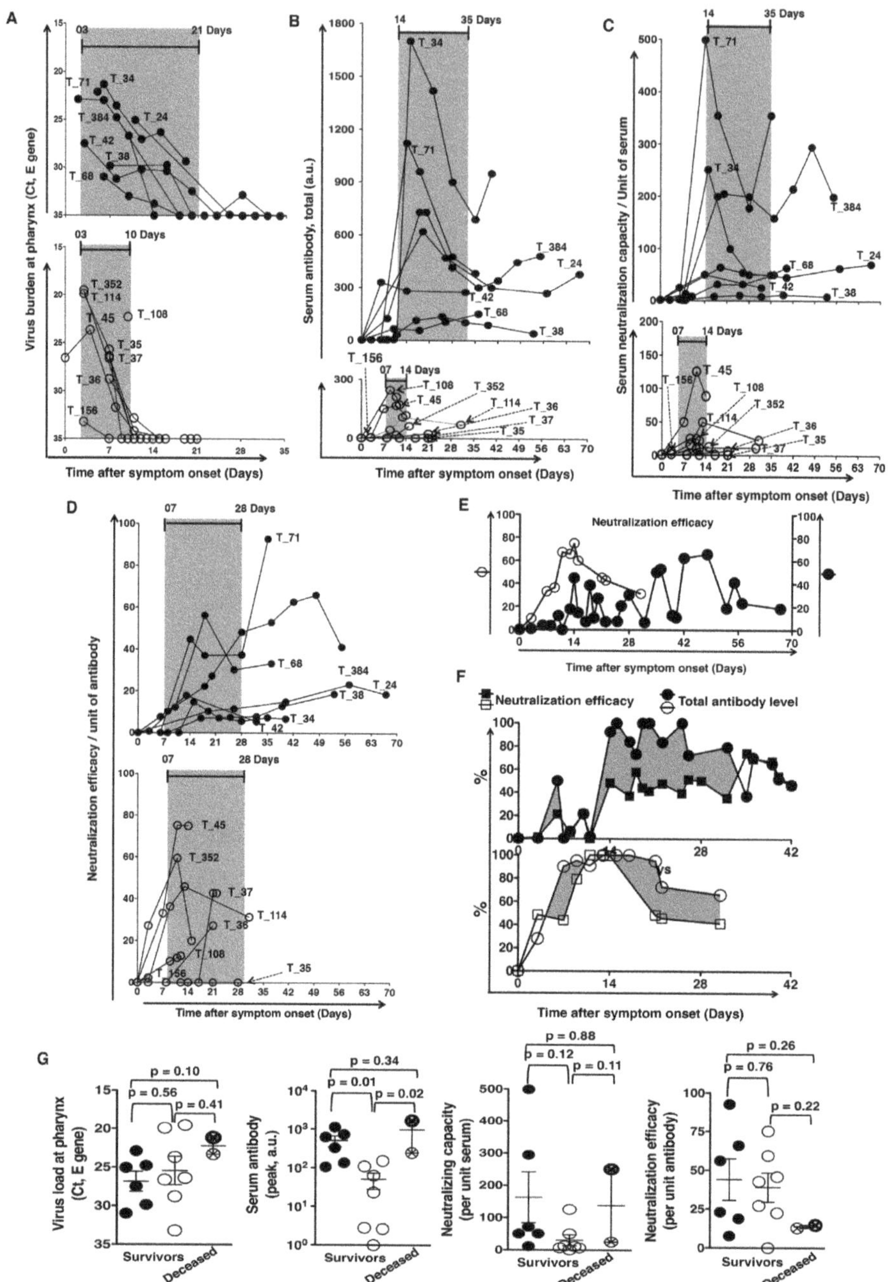

Figure 2. (**A**) *Pharyngeal virus load and clearance.* Of the participants involved, 7 patients revealed delayed clearance of virus (upper panel), and the other 8 patients eradicated the virus quickly (lower panel). (**B,C**) *Serum antibody level and neutralization titer.* (**B**) High antibody level per unit of serum and (**C**) proportionally high neutralization titer per unit of serum with persistent presence of the virus (upper panels), compared to those in the patients with rapid virus clearance

(lower panels). (**D,F**) *Neutralization efficacy per unit of antibody.* (**D**) Comparable neutralization efficacy per unit of antibody between patients with viral persistence (upper panel) and patients with rapid eradication (lower panel). (**E**). Longer average time for peak neutralization efficacy per unit of antibody in patients with viral persistence (closed circle) than those who cleared virus rapidly (open circle). (**F**) There were many more antibodies of compromised neutralization efficacy before the time of peak efficacy (shaded area) in patients with delayed clearance (upper panel), compared to those in patients with rapid clearance (lower panel). (**G**) *Viral load and antibody response in the deceased.* The deceased (the two crossed circles) had higher viral loads on presentation, higher amount of antibody and higher neutralization capacity in unit serum, but they had lower neutralization efficacy per unit of antibody, compared to those who survived (non-crossed circles). (The patients are tagged by our national serial number of COVID-19 cases).

We then used a neutralization test to evaluate the quality of antibodies with two assessments, neutralizing capacity per unit serum and neutralization efficacy per unit antibody. The capacity was higher in the delayed clearance group (168.00 ± 63.42 vs. 29.68 ± 14.82, $p = 0.0413$; Figure 2C). However, the neutralization efficacy per unit antibody was comparable between the delayed and the rapid clearance groups (41.37 ± 11.49 and 35.70 ± 8.78; $p = 0.6975$; Figure 2D). It is also interesting that the time for peak neutralization efficacy was significantly longer in the delayed clearance group (34.14 ± 7.15 vs. 12.50 ± 2.35 days PSO, $p = 0.0094$; Figure 2E). As the patients in the delayed clearance group had a huge antibody response at first, there were many more antibodies of compromised efficacy before the time of peak efficacy, shown as the shaded area, compared to the patients in the rapid clearance group (Figure 2F).

Of the participants involved, 2 of the 15 patients died. The deceased had a higher viral load at presentation and a larger amount of poor quality antibodies. Strong but poor-quality antibody response, probably associated with delayed clearance of the virus, was a factor for the less favorable clinical outcome of the disease (Figure 2G).

2.3. Inflammation and Relative Viral Persistence

Only four patients in our cohort required ICU assistance. One of them belongs to the group of rapid virus clearance. The patient had other comorbidity and succumbed suddenly after a brief hospital stay. We thus do not have serial serum samples of this patient. We also do not have serial serum samples of the patients with rapid virus clearance and mild disease symptoms, owing to their brief hospital stay and lower number of follow-up tests. Only one such patient donated blood over a follow-up period of two months. We found that the levels of inflammatory cytokines and chemokines were higher in the three patients with delayed virus clearance and severe disease than in the asymptomatic patient with rapid virus clearance. Among patients with relative virus persistence and severe disease, the deceased had high levels of inflammatory cytokines including IFN-γ, IL-17A, IL-6, LIF (leukemia inhibitory factor of IL-6 family), IL-2, IL-3, IL-16, IL-18, and M-CSF (closed squares, Figure 3A), as well as inflammatory chemokines CXCL-9, CXCL-10 (IP-10), CCL-2, and CCL-7 (closed squares, Figure 3B). The stem cell factor (SCF) has previously been linked with airway inflammation [4,5], and the level of SCF was high in serial sera samples of the deceased patient (closed squares, Figure 3C). The deceased had a high level of immune activation-associated molecule IL-2RA (CD25) as well (closed squares, Figure 3D). For the other two patients with severe disease who survived, high levels of IL-12p70, IL-13, CXCL-9, and IL-12RA were present in one of the two (closed circles, Figure 3). There were minimal levels of proinflammatory cytokines and chemokines in the other patient with a disease of less severity (closed triangles, Figure 3). The asymptomatic patient with rapid virus clearance had detectable levels of IFN-γ, IL-6, LIF, IL-16, IL-18, M-CSF, CXCL-9, SCF, and IL-12RA only on day 7 PSO during two-month follow-up (open circles, Figure 3). Interestingly, IL-10 was always higher in this patient, compared to the patients with severe disease (open circles, Figure 3E).

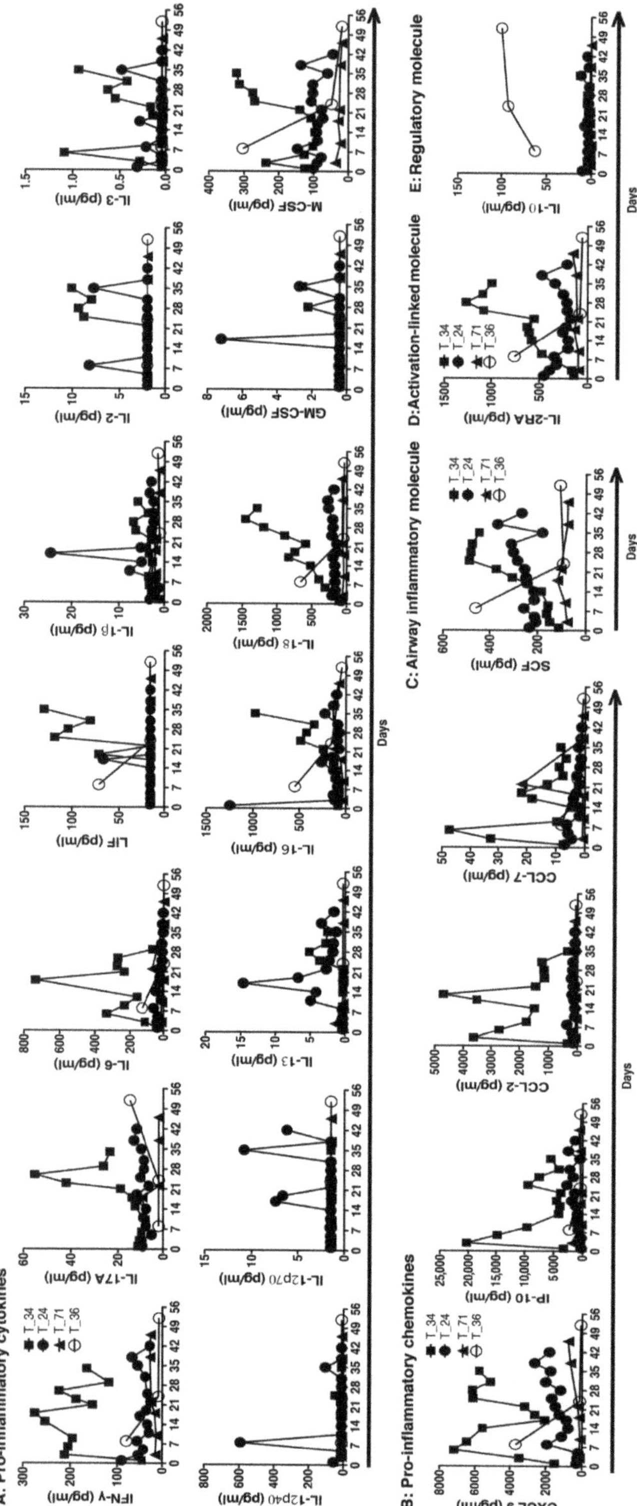

Figure 3. Cytokine profiles of RT-PCR-confirmed COVID-19 patients. Kinetics of (**A**) pro-inflammatory cytokines, (**B**) pro-inflammatory chemokines, (**C**) airway inflammatory molecule SCF, (**D**) immune activation-associated molecule IL-2RA, and (**E**) regulatory molecule IL-10 in the serial sera samples from stated patients during their disease course. Sera were collected from RT-PCR-confirmed COVID-19 patients. (The patients are tagged by our national serial number of COVID-19 cases).

3. Discussion

Our results show strong antibody response in patients with relative viral persistence at the pharynx. They had excessive antibodies of compromised quality in an early stage with the delay in peak of virus neutralization efficacy per unit of antibody. Weak antibody response of patients with rapid viral clearance was also effective, with substantial and comparable neutralization efficacy. Viral persistence boosted inflammatory immune activation. Among patients with delayed virus clearance and ICU assistance, concentrations of proinflammatory cytokines and chemokines were higher in the deceased patient than that in the patients who suffered from severe disease but eventually survived.

Strong antibody response, in terms of high antibody level and proportionally high neutralization titer in the sera, with slower clearance of the SARS-CoV-2 virus has been reported in the literature [6–9]. The antibody levels began to decline two weeks PSO [10,11]. We also observed strong antibody responses in our patients with relative virus persistence, and the antibody levels started to decline two weeks PSO. Despite this decline in antibody level, we found that neutralization efficacy per unit of antibody remained the same or continued to increase in these patients. This indicates that the proportion of antibodies with lower neutralization efficacy gradually decreases, while the proportion of higher efficacy gradually increases with time. The phenomenon of neutralization efficacy increasing over time is in line with the known maturation process of the antibody response. Created through random VDJ recombination, the B cell receptor (BCR) repertoire is highly heterogeneous. Clonal selection is achieved through stimulation and response where B cells with BCR and antibodies of effective neutralization ability gradually expand and become the major B cell pool responding to the virus. Neutralization represents the antibody's ability to protect against specific pathogens. It deserves special attention because there is a population of antibodies with poor neutralization capacity in the early stages of the antibody reaction. One of the most concerning risks of convalescent plasma therapy for COVID-19 is that some plasma antibodies may in fact not be protective [12]. They could even be harmful due to mechanisms such as antibody-dependent enhancement (ADE) [13]. Therefore, we must be cautious about the timing of plasma procurement from patients who have recovered from the illness.

People tend to try to link the association between the amounts of virus in respiratory samples and the severity of illness [14,15]. However, the persistent presence of the virus rather than the absolute amount of virus at the throat was responsible for a strong and early antibody response in our cohort of COVID-19 patients. A strong and early antibody response likely predominantly comprises less protective and potentially even deleterious antibodies. In patients with SARS, it was reported that poor clinical outcomes were associated with the early appearance of antibodies [16]. Patients with difficulty eradicating the virus suffer from the damage caused by both the virus and the ineffective potentially deleterious antibodies. In our study, patients with viral persistence and an earlier and stronger antibody response tended to be older. This may explain the vulnerability to COVID-19 in the elderly.

In our observations, low or even no detectable antibodies did not necessarily represent an absence of immunity. Although the absolute antibody quantity in these patients is low, the neutralization efficacy per unit of antibody is equivalent to that of the group with higher antibody levels, indicating that patients with low antibody quantities also have a considerable number of mature B cells secreting effective antibodies. Upon subsequent encounters with the virus, these B cells will likely expand with a memory response and may produce effective antibodies in quantities sufficient to protect the host.

More and more evidence indicates a hyperinflammatory response to SARS-CoV-2 contributes to the development of ARDS, disease severity, and death in COVID-19 [17–22]. We also detected high levels of proinflammatory cytokines and chemokines in patients with viral persistence and severe disease requiring intensive care. This trend was much more exaggerated in the deceased. The deceased patient had elevated IL-6 with an increase in other cytokine and chemokine levels in the serum, also similar to the reported literature [19,23–25]. There was persistence of elevated levels of IFN-γ IL-17A, IL-6 family member LIF, and many other pro-inflammatory cytokines and chemokines in this deceased patient, compared to those in patients who had delayed virus clearance and suffered from severe disease but eventually survived. Interestingly, there was a high IL-10 level maintained in the asymptomatic patient that may contribute to curtailed disease severity.

We understand that the small cohort size is the limitation of our study. The kinetics of viral persistence, antibody response, and cytokine profile we observed in only 15 COVID-19 patients in our study were parallel to the literature. However, our analysis revealed that even though antibody levels begin to decline two weeks PSO, the neutralization efficacy per unit of antibody remained the same or continued to increase. The process of antibody maturation was delayed in patients with virus persistence. This indicates that the population of antibodies with poor neutralization capacity in the early stages may be deleterious instead of helpful. Timing of plasma procurement can be a critical factor for convalescent plasma therapy. Our results also suggest management of proinflammatory cytokines other than IL-6 may help toward recovery from severe COVID-19, as evidenced by the consolidated benefit of low-dose corticosteroid in treatment [26].

4. Methods

4.1. Patients, Sample Collection and Handling, Biobanking, and Ethics Statement

All COVID-19 patients were RT-PCR-confirmed and placed in negative pressure isolation rooms in our hospital. Nasopharyngeal or oropharyngeal throat swab specimens were collected on stated days post symptom onset for serial RT-PCR tracking. Pharyngeal specimens were also used to isolate and culture SARS-CoV-2 virus strains. One virus strain was used for a neutralization antibody test in our BSL-3 facility. Antibody tests were carried out using serum samples. Peripheral blood was collected for routine medical tests, and serum samples remaining after routine medical tests were used in our research.

Isolated virus strains are deposited in our institutional depository. Sequences of the virus strains are also deposited in the depository of the Taiwan Centers for Disease Control (Taiwan CDC).

This research was performed with informed consent from patients or their families. Specimen sampling and transportation were handled according to the criteria of the Taiwan CDC. This study was approved by the Institutional Review Board of Chang Gung Medical Foundation, Linkou Medical Center, Taoyuan City, Taiwan.

4.2. SARS-CoV-2 Nucleic Acid Detection

Nasopharyngeal or oropharyngeal throat swab specimens were collected from patients. Test for SARS-CoV-2 nucleic acid followed standard protocols. RNA was extracted from clinical samples with the LabTurbo system (Taigen, Taiwan). A 25 μL reaction contained 5 μL of RNA, 12.5 μL of 2 × reaction buffer provided with the Superscript III one-step RT-PCR system with Platinum Taq Polymerase (AgPath-ID One-step RT-PCR Kit), 1 μL of reverse transcriptase/Taq mixture from the kit, 0.4 μL of a 50 mM magnesium sulfate solution (Invitrogen), and 1 μg of nonacetylated bovine serum albumin (Roche). All oligonucleotides were synthesized and provided by Tib-Molbiol (Berlin, Germany). Thermal cycling was performed at 48 °C for 30 min for reverse transcription, followed by 95 °C for 10 min and then 45 cycles of 95 °C for 10 s, 65 °C for the 30 s [27].

4.3. COVID-19 Serum Antibody Detection

To evaluate the antibody response, the levels of total IgG in patients' sera were semi-quantified by ELISA (cat No. WS-1096, WANTAI SARS-CoV-2 Ab ELISA, China) through the use of a Triturus ELISA processor, following manufacturer's instructions. WANTAI SARS-CoV-2 Ab ELISA is a two-step incubation antigen "sandwich" enzyme immunoassay kit, which uses polystyrene microwell strips pre-coated with recombinant SARS-CoV-2 antigen. The results had previously been verified with another ELISA kit (Anti-SARS-CoV-

2 ELISA IgG, Euroimmun, Germany), and the ELISA was done in accordance with the manufacturer's instructions.

4.4. Neutralization Antibody Test (NAT)

The neutralizing antibody test of COVID-19 followed the standard protocol of a plaque reduction neutralization test. Vero cells were regularly maintained in minimal essential medium (MEM) supplemented with 10% (v/v) fetal bovine serum (FBS). COVID-19 virus was propagated in Vero cells in a maintenance medium consisting of MEM supplemented with 0% FBS. Serum samples were inactivated at 56 °C for 30 min before use. Serial two-fold dilutions of sera were mixed with an equal volume of COVID-19 virus suspension containing 100 × the median tissue culture infectious dose ($TCID_{50}$). The mixture was incubated for 2 h at 37 °C, and then an equal volume of suspended VeroE6 cells (approximately 30,000 cells/well) was added to each well. Following incubation for 1 week at 37 °C, cells were fixed with 5% glutaraldehyde and stained with 0.1% crystal violet. Serum neutralization titers were calculated and expressed as the reciprocals of the highest serum dilution that inhibits cytopathic effects.

4.5. COVID-19 Serum Cytokine and Chemokine Detection

To evaluate the response of cytokines, chemokines, and other immune molecules, sera were quantified by Bio-Plex Pro Human Cytokine 48-Plex Screening kit (Bio-Rad), as per the manufacturer's instructions.

Author Contributions: Conceptualization, A.D., C.-T.H. and C.-T.Y.; methodology, C.-G.H. and K.-C.T.; validation, S.-R.S., C.-T.H. and C.-T.Y.; formal analysis, A.D. and C.-G.H.; investigation, S.-M.L., P.-Y.C., M.-J.H. and Y.-C.H.; data curation, A.D. and C.-G.H.; writing—original draft preparation, A.D. and C.-T.H.; writing—review and editing, A.D., C.-T.H. and C.-T.Y.; supervision, C.-T.H., S.-R.S. and C.-T.Y.; project administration, C.-T.H., S.-R.S. and C.-T.Y.; funding acquisition, A.D., C.-T.H., S.-R.S. and C.-T.Y. All authors have read and agreed to the published version of the manuscript.

Funding: Avijit Dutta was supported by special researcher recruitment program of the Ministry of Science and Technology, Taiwan (MOST-107-2811-B-182A-505 & MOST 108-2811-B-182A-517). The Research Center for Emerging Viral Infections receives support from The Featured Areas Research Center Program within the framework of the Higher Education Sprout Project by the Ministry of Education (MOE), Taiwan. This work was also supported by the Projects CMRPG3J1501 (C.-T.H.) and CORPG5K0011 (C.-T.Y.) from Medical Research Project Fund, Chang Gung Memorial Hospital, Taiwan.

Institutional Review Board Statement: The study was conducted according to the guidelines of the Declaration of Helsinki, and approved by the Institutional Review Board of Chang Gung Memorial Hospital, Taoyuan City, Taiwan.

Informed Consent Statement: Informed consent was obtained from all subjects involved in the study.

Conflicts of Interest: The authors declare no competing interest.

Statistical Analysis: We used Graph Pad Prism version 5 for statistical analyses. Data represented as mean $\pm$ SD, Pearson r values for correlation, and p values for two-tailed unpaired Student's t-test in the text.

References

1. Guo, L.; Ren, L.; Yang, S.; Xiao, M.; Chang, D.; Yang, F.; Dela Cruz, C.S.; Wang, Y.; Wu, C.; Xiao, Y.; et al. Profiling early humoral response to diagnose novel coronavirus disease (COVID-19). *Clin. Infect. Dis.* **2020**, ciaa310. [CrossRef]
2. Ong, D.S.Y.; de Man, S.J.; Lindeboom, F.A.; Koeleman, J.G.M. Comparison of diagnostic accuracies of rapid serological tests and ELISA to molecular diagnostics in patients with suspected coronavirus disease 2019 presenting to the hospital. *Clin. Microbiol. Infect.* **2020**, S1198. [CrossRef]
3. Beavis, K.G.; Matushek, S.M.; Abeleda, A.P.F.; Bethel, C.; Hunt, C.; Gillen, S.; Moran, A.; Tesic, V. Evaluation of the EUROIMMUN anti-SARS-CoV-2 ELISA Assay for detection of IgA and IgG antibodies. *J. Clin. Virol.* **2020**, *129*, 104468. [CrossRef] [PubMed]

4. Lukacs, N.W.; Strieter, R.M.; Lincoln, P.M.; Brownell, E.; Pullen, D.M.; Schock, H.J.; Chensue, S.W.; Taub, D.D.; Kunkel, S.L. Stem cell factor (c-kit ligand) influences eosinophil recruitment and histamine levels in allergic airway inflammation. *J. Immunol.* **1996**, *156*, 3945–3951.

5. Derakhshan, T.; Samuchiwal, S.K.; Hallen, N.; Bankova, L.G.; Boyce, J.A.; Barrett, N.A.; Austen, K.F.; Dwyer, D.F. Lineage-specific regulation of inducible and constitutive mast cells in allergic airway inflammation. *J. Exp. Med.* **2021**, *218*, e20200321. [CrossRef]

6. Zhang, X.; Lu, S.; Li, H.; Wang, Y.; Lu, Z.; Liu, Z.; Lai, Q.; Ji, Y.; Huang, X.; Li, Y.; et al. Viral and antibody kinetics of COVID-19 patients with different disease severities in acute and convalescent phases: A 6-month follow-up study. *Virol. Sin.* **2020**, *35*, 820–829. [CrossRef]

7. Sun, J.; Tang, X.; Bai, R.; Liang, C.; Zeng, L.; Lin, H.; Yuan, R.; Zhou, P.; Huang, X.; Xiong, Q.; et al. The kinetics of viral load and antibodies to SARS-CoV-2. *Clin. Microbiol. Infect.* **2020**, *26*, 1690.e1. [CrossRef]

8. Wang, Y.; Zhang, L.; Sang, L.; Ye, F.; Ruan, S.; Zhong, B.; Song, T.; Alshukairi, A.N.; Chen, R.; Zhang, Z.; et al. Kinetics of viral load and antibody response in relation to COVID-19 severity. *J. Clin. Investig.* **2020**, *130*, 5235–5244. [CrossRef]

9. Marklund, E.; Leach, S.; Axelsson, H.; Nyström, K.; Norder, H.; Bemark, M.; Angeletti, D.; Lundgren, A.; Nilsson, S.; Andersson, L.-M.; et al. Serum-IgG responses to SARS-CoV-2 after mild and severe COVID- 19 infection and analysis of IgG non-responders. *PLoS ONE* **2020**, *15*, e0241104. [CrossRef] [PubMed]

10. Seow, J.; Graham, C.; Merrick, B.; Acors, S.; Pickering, S.; Steel, K.J.A.; Hemmings, O.; O'Byrne, A.; Kouphou, N.; Galao, R.P.; et al. Longitudinal observation and decline of neutralizing antibody responses in the three months following SARS-CoV-2 infection in humans. *Nat. Microbiol.* **2020**, *5*, 1598–1607. [CrossRef]

11. Lau, E.H.Y.; Tsang, O.T.Y.; Hui, D.S.C.; Kwan, M.Y.W.; Chan, W.-H.; Chiu, S.S.; Ko, R.L.W.; Chan, K.H.; Cheng, S.M.S.; Perera, R.A.P.M.; et al. Neutralizing antibody titres in SARS-CoV-2 infections. *Nat. Commun.* **2021**, *12*, 63. [CrossRef]

12. Robbiani, D.F.; Gaebler, C.; Muecksch, F.; Lorenzi, J.C.C.; Wang, Z.; Cho, A.; Agudelo, M.; Barnes, C.O.; Gazumyan, A.; Finkin, S.; et al. Convergent antibody responses to SARS-CoV-2 in convalescent individuals. *Nature* **2020**, *584*, 437–442. [CrossRef]

13. Wang, S.F.; Tseng, S.P.; Yen, C.H.; Yang, J.Y.; Tsao, C.H.; Shen, C.W.; Chen, K.H.; Liu, F.T.; Liu, W.T.; Chen, Y.M.; et al. Antibody-dependent SARS coronavirus infection is mediated by antibodies against spike proteins. *Biochem. Biophys. Res. Commun.* **2014**, *451*, 208–214. [CrossRef]

14. Sun, J.; Xiao, J.; Sun, R.; Tang, X.; Liang, C.; Lin, H.; Zeng, L.; Hu, J.; Yuan, R.; Zhou, P.; et al. Prolonged persistence of SARS-CoV-2 RNA in body fluids. *Emerg. Infect. Dis.* **2020**, *26*, 1834–1838. [CrossRef]

15. Carmo, A.; Pereira-Vaz, J.; Mota, V.; Mendes, A.; Morais, C.; Coelho da Silva, A.; Camilo, E.; Silva Pinto, C.; Cunha, E.; Pereira, J.; et al. Clearance and persistence of SARS-CoV-2 RNA in patients with COVID-19. *J. Med. Virol.* **2020**, *92*, 2227–2231. [CrossRef] [PubMed]

16. Ho, M.S.; Chen, W.J.; Chen, W.J.; Lin, S.F.; Wang, M.C.; Di, J.; Lu, Y.T.; Liu, C.L.; Chang, S.C.; Chao, C.L.; et al. Neutralizing antibody response and SARS severity. *Emerg. Infect. Dis.* **2005**, *11*, 1730–1737. [CrossRef]

17. Wu, C.; Chen, X.; Cai, Y.; Xia, J.; Zhou, X.; Xu, S.; Huang, H.; Zhang, L.; Zhou, X.; Du, C.; et al. Risk factors associated with acute respiratory distress syndrome and death in patients with Coronavirus disease 2019 pneumonia in Wuhan, China. *JAMA Intern. Med.* **2020**, *180*, 934–943. [CrossRef] [PubMed]

18. Mehta, P.; McAuley, D.F.; Brown, M.; Sanchez, E.; Tattersall, R.S.; Manson, J.J. COVID-19: Consider cytokine storm syndromes and immunosuppression. *Lancet* **2020**, *395*, 1033–1034. [CrossRef]

19. Del Valle, D.M.; Kim-Schulze, S.; Hsin-Hui, H.; Beckmann, N.D.; Nirenberg, S.; Wang, B.; Lavin, Y.; Swartz, T.; Madduri, D.; Stock, A.; et al. An inflammatory cytokine signature helps predict COVID-19 severity and survival. *Nat. Med.* **2020**, *26*, 1636–1643. [CrossRef] [PubMed]

20. Ong, E.Z.; Chan, Y.F.Z.; Leong, W.Y.; Lee, N.M.Y.; Kalimuddin, S.; Mohideen, S.M.H.; Chan, K.S.; Tan, A.T.; Bertoletti, A.; Ooi, E.E.; et al. A dynamic immune response shapes COVID-19 progression. *Cell Host Microbe* **2020**, *27*, 879–882. [CrossRef]

21. Blanco-Melo, D.; Nilsson-Payant, B.E.; Liu, W.C.; Uhl, S.; Hoagland, D.; Møller, R.; Jordan, T.X.; Oishi, K.; Panis, M.; Sachs, D.; et al. Imbalanced host response to SARS-CoV-2 drives development of COVID-19. *Cell* **2020**, *181*, 1036–1045. [CrossRef] [PubMed]

22. Wheatley, A.K.; Juno, J.A.; Wang, J.J.; Selva, K.J.; Reynaldi, A.; Tan, H.X.; Lee, W.S.; Wragg, K.M.; Kelly, H.G.; Esterbauer, R.; et al. Evolution of immune responses to SARS-CoV-2 in mild-moderate COVID-19. *Nat. Commun.* **2021**, *12*, 1162. [CrossRef] [PubMed]

23. Broman, N.; Rantasärkkä, K.; Feuth, T.; Valtonen, M.; Waris, M.; Hohenthal, U.; Rintala, E.; Karlsson, A.; Marttila, H.; Peltola, V.; et al. IL-6 and other biomarkers as predictors of severity in COVID-19. *Ann. Med.* **2021**, *53*, 410–412. [CrossRef]

24. Ruan, Q.; Yang, K.; Wang, W.; Jiang, L.; Song, J. Clinical predictors of mortality due to COVID-19 based on an analysis of data of 150 patients from Wuhan, China. *Intensive Care Med.* **2020**, *46*, 846–848. [CrossRef]

25. Chen, G.; Wu, D.; Guo, W.; Cao, Y.; Huang, D.; Wang, H.; Wang, T.; Zhang, X.; Chen, H.; Yu, H.; et al. Clinical and immunological features of severe and moderate Coronavirus disease 2019. *J. Clin. Investig.* **2020**, *130*, 2620–2629. [CrossRef]

26. Horby, P.; Lim, W.S.; Emberson, J.R.; Mafham, M.; Bell, J.L.; Linsell, L.; Staplin, N.; Brightling, C.; Ustianowski, A.; RECOVERY Collaborative Group. Dexamethasone in hospitalized patients with Covid-19. *N. Engl. J. Med.* **2021**, *384*, 693–704. [PubMed]

27. Corman, V.M.; Landt, O.; Kaiser, M.; Molenkamp, R.; Meijer, A.; Chu, D.K.; Bleicker, T.; Brünink, S.; Schneider, J.; Schmidt, M.L.; et al. Detection of 2019 novel coronavirus (2019-nCoV) by real-time RT-PCR. *Euro Surveill.* **2020**, *25*, 2000045. [CrossRef]

pathogens

Article

Longitudinal Development of Antibody Responses in COVID-19 Patients of Different Severity with ELISA, Peptide, and Glycan Arrays: An Immunological Case Series

Jasmin Heidepriem [1,†], Christine Dahlke [2,3,4,*,†], Robin Kobbe [2], René Santer [5], Till Koch [2,3,4], Anahita Fathi [2,3,4], Bruna M. S. Seco [1], My L. Ly [2,3,4], Stefan Schmiedel [2], Dorothee Schwinge [6], Sonia Serna [7], Katrin Sellrie [1], Niels-Christian Reichardt [7,8], Peter H. Seeberger [1], Marylyn M. Addo [2,3,4,*], Felix F. Loeffler [1,*] and on behalf of the ID-UKE COVID-19 Study Group [‡]

1. Department of Biomolecular Systems, Max Planck Institute of Colloids and Interfaces, Am Muehlenberg 1, 14476 Potsdam, Germany; jasmin.heidepriem@mpikg.mpg.de (J.H.); BrunaMara.SilvaSeco@mpikg.mpg.de (B.M.S.S.); Katrin.Sellrie@mpikg.mpg.de (K.S.); Peter.Seeberger@mpikg.mpg.de (P.H.S.)
2. Division of Infectious Diseases, First Department of Medicine, University Medical Center Hamburg-Eppendorf, 20251 Hamburg, Germany; r.kobbe@uke.de (R.K.); t.koch@uke.de (T.K.); a.fathi@uke.de (A.F.); m.ly@uke.de (M.L.L.); s.schmiedel@uke.de (S.S.)
3. Department of Clinical Immunology of Infectious Diseases, Bernhard Nocht Institute for Tropical Medicine, 20251 Hamburg, Germany
4. German Center for Infection Research, Partner Site Hamburg-Lübeck-Borstel-Riems, 20251 Hamburg, Germany
5. Department of Pediatrics, University Medical Center Hamburg-Eppendorf, 20251 Hamburg, Germany; r.santer@uke.de
6. I. Department of Medicine, University Medical Center Hamburg-Eppendorf, 20251 Hamburg, Germany; dschwing@uke.de
7. Glycotechnology Laboratory, Center for Cooperative Research in Biomaterials (CIC biomaGUNE), Basque Research and Technology Alliance (BRTA), Paseo de Miramon 182, 20014 Donostia San Sebastián, Spain; sserna@cicbiomagune.es (S.S.); nreichardt@cicbiomagune.es (N.-C.R.)
8. CIBER-BBN, Paseo Miramón 182, 20014 San Sebastián, Spain
* Correspondence: c.dahlke@uke.de (C.D.); m.addo@uke.de (M.M.A.); felix.loeffler@mpikg.mpg.de (F.F.L.)
† These authors contributed equally.
‡ Members of the ID-UKE COVID-19 Study Group are listed at the end of the manuscript.

Citation: Heidepriem, J.; Dahlke, C.; Kobbe, R.; Santer, R.; Koch, T.; Fathi, A.; Seco, B.M.S.; Ly, M.L.; Schmiedel, S.; Schwinge, D.; et al. Longitudinal Development of Antibody Responses in COVID-19 Patients of Different Severity with ELISA, Peptide, and Glycan Arrays: An Immunological Case Series. *Pathogens* **2021**, *10*, 438. https://doi.org/10.3390/pathogens10040438

Academic Editors: Philipp A. Ilinykh and Kai Huang

Received: 2 March 2021
Accepted: 1 April 2021
Published: 6 April 2021

Publisher's Note: MDPI stays neutral with regard to jurisdictional claims in published maps and institutional affiliations.

Abstract: The current COVID-19 pandemic is caused by the severe acute respiratory syndrome coronavirus-2 (SARS-CoV-2). A better understanding of its immunogenicity can be important for the development of improved diagnostics, therapeutics, and vaccines. Here, we report the longitudinal analysis of three COVID-19 patients with moderate (#1) and mild disease (#2 and #3). Antibody serum responses were analyzed using spike glycoprotein enzyme linked immunosorbent assay (ELISA), full-proteome peptide, and glycan microarrays. ELISA immunoglobulin A, G, and M (IgA, IgG, and IgM) signals increased over time for individuals #1 and #2, whereas #3 only showed no clear positive IgG and IgM result. In contrast, peptide microarrays showed increasing IgA/G signal intensity and epitope spread only in the moderate patient #1 over time, whereas early but transient IgA and stable IgG responses were observed in the two mild cases #2 and #3. Glycan arrays showed an interaction of antibodies to fragments of high-mannose and core N-glycans, present on the viral shield. In contrast to protein ELISA, microarrays allow for a deeper understanding of IgA, IgG, and IgM antibody responses to specific epitopes of the whole proteome and glycans of SARS-CoV-2 in parallel. In the future, this may help to better understand and to monitor vaccination programs and monoclonal antibodies as therapeutics.

Keywords: SARS-CoV-2; COVID-19; full proteome; peptide microarrays; glycan microarrays

1. Introduction

The novel severe acute respiratory syndrome coronavirus 2 (SARS-CoV-2) was first described in Wuhan, China, in January 2020, as the causative agent of COVID-19 [1]. CoVs were not considered to be highly pathogenic, until the emergence of SARS-CoV [2–4] in 2002, and the Middle East respiratory syndrome (MERS)-CoV in 2012 [5]. With SARS-CoV-2, three CoVs have passed the species barriers from animal to human in the last 20 years, causing severe respiratory diseases. Based on their pathogenic and epidemic potential, the World Health Organization (WHO) has classified all three CoVs as priority pathogens to accelerate the development of vaccines and therapeutics to prevent epidemics.

The world is still confronted with the SARS-CoV-2 pandemic. This virus belongs to the *Betacoronavirus* genus of the Coronaviridae family and has genetic similarity with SARS-CoV. Within about one year, more than 114 million people have been infected globally, with more than 2.5 million reported deaths as of 2 March 2021 [6]. The infection presents with different symptoms and a wide spectrum of severity [7]. Some patients only experience very mild symptoms like a cough, while others show a very severe form of the disease that leads to bilateral pneumonia.

An efficient countermeasure to limit an outbreak includes specific and sensitive diagnostics. Polymerase chain reaction (PCR) is used to measure SARS-CoV-2 particles, whereas antibodies are measured by enzyme linked immunosorbent assay (ELISA), the gold standard for the detection of SARS-CoV-2 specific antibodies. These tests mainly rely on the binding of serum antibodies to the SARS-CoV-2 spike glycoprotein (S) [8]. The advantage of ELISAs is their simplicity and standardized protocol. A disadvantage is the limitation in sensitivity and specificity, since they lack information on specific epitopes.

Array technologies can help to fill this gap and identify epitopes that are targeted by antibodies, which in turn may be used to support the development of vaccines or monoclonal antibodies as therapeutics. High-density peptide arrays enable the rapid identification of antigen epitopes recognized by antibodies for many applications [9]. Pathogen-specific peptide arrays help to identify biomarkers for (early) detection of diseases [10]. Glycan arrays allow for the characterization and surveillance of viruses, identification of biomarkers, profiling of immune responses to vaccines, and epitope mapping [11,12].

In this study, we evaluate three distinct assays to identify the development of SARS-CoV-2 specific antibodies: (i) peptide arrays, covering the whole SARS-CoV-2 proteome as overlapping linear peptides, (ii) glycan arrays with a selected glycan library, and (iii) spike glycoprotein ELISA. We assess the ability of these assays to identify distinct epitopes, which can serve as potential biomarkers for disease progression. In combination with the clinical data of patients, we gained insights into immunoglobulin A, G, and M (IgA, IgG, and IgM) responses during COVID-19 progression.

Here, we report longitudinal antibody response data from three SARS-CoV-2-positive patients, sampled three times. While patient #1 had a moderate course of disease and was hospitalized (no ventilation), patient #2 experienced mild symptoms. Patient #3, who also had mild symptoms, was sampled only twice during disease and once 180 days before infection, which served as the negative control. Finally, for comparison, we added one sample of a single time point from another COVID-19 patient #4.

2. Materials and Methods

2.1. Patient Material

Blood samples were collected at the University Medical Center Hamburg-Eppendorf and the serum was immediately separated at 2000 g for 10 min, aliquoted, frozen, and stored at −80 °C. Table 1 lists information on patients and blood collection days.

Table 1. Patient and serum sample information; samples analyzed with peptide and glycan microarrays.

Sample	Patient ID	Gender	Age [y]	Symptoms	Hospitalized	Day of Serum Collection after Onset of Symptoms
1						d6
2	#1	Male	64	Moderate	Yes	d10
3						d22
4						d3
5	#2	Female	62	Mild	No	d15
6						d24
7						d-180
8	#3	Male	37	Mild	No	d4
9						d11
10	#4	Female	23	Mild	No	d12

2.2. Serum IgA, IgG, and IgM Elisa

Semi-quantitative SARS-CoV-2 IgA, IgG, and IgM enzyme-linked immunosorbent assay (ELISA) targeting the S1-Domain of the S-spike protein subunit were performed (Euroimmun AG, Lübeck, Germany) according to the manufacturer's instructions. Optical density was determined at a wavelength of 450 nm (OD450) and correction set to 620 nm. Ratios were calculated by Ratio = (Extinction control or sample)/(Extinction calibrator). A calibrator and positive control were provided with each ELISA kit. According to the manufacturer, a ratio of ≥ 1.1 should be regarded as positive and the manufacturer reports a specificity of 92.5 % for IgA, 99.3 % for IgG, and 98.6 % for IgM.

2.3. Peptide and Glycan Microarrays

The whole proteome of SARS-CoV-2 (GenBank ID: MN908947.3) was mapped as 4883 spots of overlapping 15-mer peptides with a lateral shift of two AA on peptide microarrays, obtained from PEPperPRINT GmbH (Heidelberg, Germany). Glycan microarrays containing a selection of 135 glycans were produced at CIC biomaGUNE (San Sebastián, Spain) [13]. Patient sera were diluted 1:200 (peptide) or 1:100 (glycan) and incubated on the arrays overnight. Afterwards, IgG, IgM, and IgA serum antibody interactions were differentially detected with fluorescently labeled secondary antibodies. For details, see Supplementary Materials.

3. Results

We collected blood of COVID-19 patients at different time points (Table 1) and used ELISA, peptide, and glycan microarrays to evaluate the kinetics of antibody development in detail.

Patient #1, a 64-year-old male, developed general weakness, myalgia and headache, intermittent episodes of very high fever, and subsequently, a productive cough. Two days after the first symptoms, he was tested positive for SARS-CoV-2 by RT-PCR. At that time point, the fever had already subsided, but a low-grade temperature recurred in the second week. An increase of C-reactive protein (53 mg/dL) required oral treatment with beta-lactamase antibiotic. The patient was hospitalized for four days and showed moderate but typical ground glass opacities on a high-resolution thorax computed tomography scan; he fully recovered without ventilation support. The patient did not require intensive care treatment or ventilation and the symptoms were moderate, due to hospitalization. Patient #2, his wife, a 62-year-old female, tested SARS-CoV-2 positive six days after her husband's first symptoms. She had high viral shedding of SARS-CoV-2 monitored by RT-PCR, although she reported only very mild clinical symptoms of COVID-19, such as sub-febrile temperatures, a mild cough, and a constant sense of well-being, as stated by Pfefferle and colleagues [14]. Patient #3 tested positive for SARS-CoV-2 with a mild course of disease, without hospitalization. Since this participant donated serum on a regular basis, a serum sample was collected 180 days before the emergence of SARS-CoV-2, serving as a negative control. In addition, one sample served as another SARS-COV-2 positive control (#4, single time point d12, mild symptoms, see Supplementary Materials).

To evaluate the kinetics of B-cell epitopes during the mild and moderate courses of COVID-19, we first performed ELISA (EUROIMMUN, Lübeck, Germany) analysis (Figure 1, Supplementary A Table S1). This test relies on the S1 fragment of the spike glycoprotein (commercial test, likely AA1–685 of spike protein with glycosylation pattern). The IgG and IgM signals of patients #1 and #2 were below the threshold at early time points on day 6 (d6) and d3 respectively. The IgA signal of patient #1 was already highly positive on day 6 and increased until day 22. Patient #1 showed a positive signal for IgG only on day 22. The earliest positive results in patient #2 for IgG and IgM signals were measured on d15 and were also positive on d24, but only the IgG signal increased further over time. The IgA level of patient #2 showed a strong increase from the early time point d3 with an intermediate signal to the highest measured value on day 15. In patient #3, the assay failed to detect a clear positive IgG and IgM response, showing IgG signals in the intermediate level on days 4 and 32, while IgA appeared positive in all samples d-180, d4, d11 and d32 (for patient #4 see Supplementary Materials).

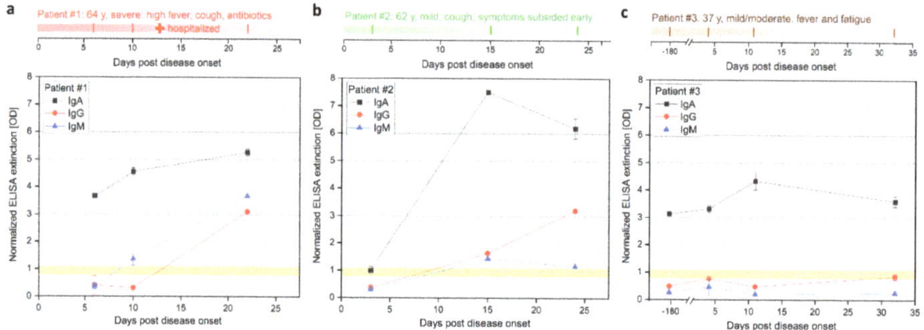

Figure 1. Longitudinal IgA, IgG, and IgM ELISA antibody response during COVID-19 disease progression in three patients. (**a**) Patients #1 (moderate), (**b**) #2 (mild), and (**c**) #3 (mild case). Patient #3 was also sampled 180 days prior infection, serving as a control. Sample from #3 collected at d32 was only analyzed with ELISA. ELISA (Euroimmune) was performed with S GP subunit 1 for detection of IgA and IgG at different days after onset of symptoms. Positive signal >1.1, negative signal <0.8, and intermediate 0.8–1.1 (highlighted yellow).

Next, we applied full-proteome peptide microarrays (see Supplementary Materials B for complete peptide microarray data). Before we evaluated the SARS-CoV-2 specific signals, an antibody threshold signal had to be established. Here, we used the serum sample from donor patient #3 (healthy negative control) 180 days before SARS-CoV-2 infection as a negative control and defined the 99.9 th percentile fluorescence intensity (i.e., 5 out of 4883 signals considered false positive) as a threshold for positive IgA- and IgG-reactive peptides (IgA: 347.8 arbitrary fluorescence units (AFU) or 6.54 transformed AFU (tAFU); IgG: 1081.4 AFU or 7.68 tAFU). Our threshold selection successfully limited the amount of presented data for intelligibility, without losing precision, as we could confirm previously published epitopes (see Discussion).

As a general trend, we observed SARS-CoV-2 protein-specific IgA and IgG responses with few defined signals, while IgM showed more signals, but without a clear trend. Thus, we focused on IgA and IgG responses. The evolution of IgA and IgG antibodies, targeting peptides of the SARS-CoV-2 proteome in patients #1 (Figure 2a,b) and #2 (Figure 2c,d) showed different dynamics: The moderate case (patient #1) showed a strong increase in IgA- and IgG-reactive peptides (above the control sample threshold) over time and eventually targeting many more epitopes (Supplementary Materials A Table S2). In comparison, the mild cases (patients #2 and #3) had a higher number of IgG- and an even higher number of IgA-reactive SARS-CoV-2 peptides already at d3 and d4, respectively, post onset of symptoms, which decreased over time (Supplementary Materials A Table S2). At these early time points, we already detected IgA and IgG-specific epitopes in the spike protein in patients #2 and #3, while patient #1 developed a high number of antibodies targeting spike epitopes only later at d22.

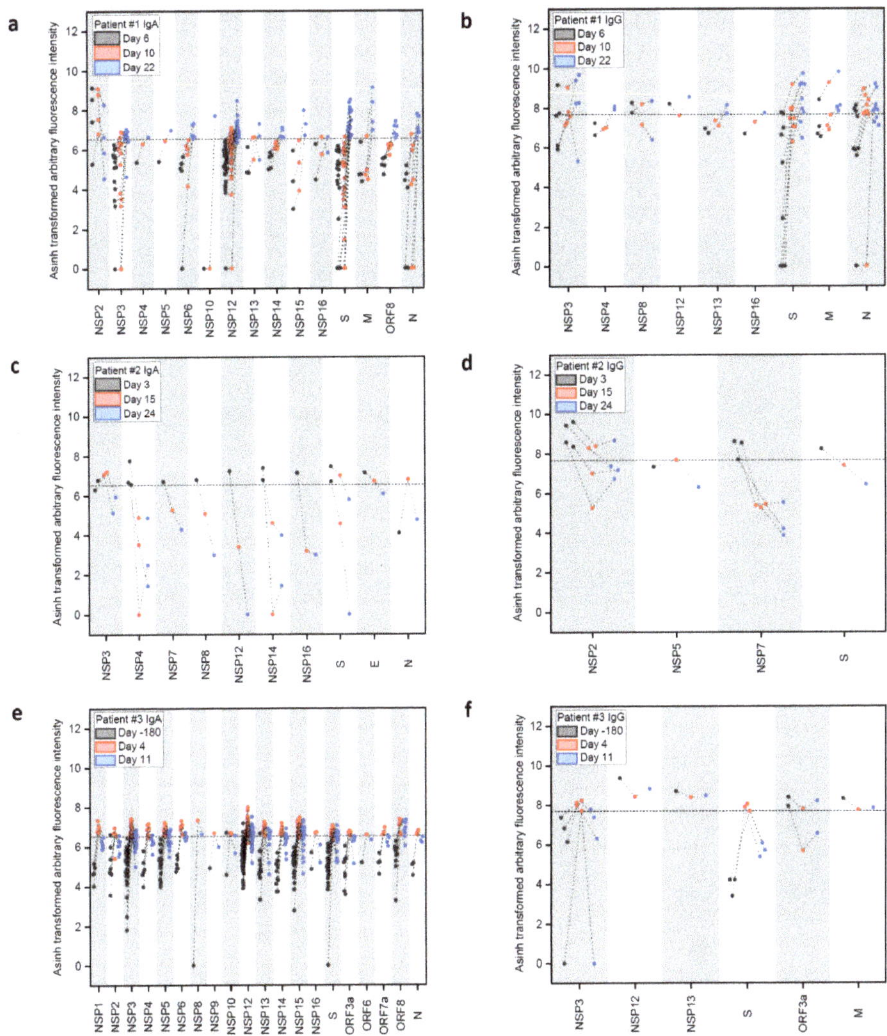

Figure 2. Longitudinal IgA and IgG antibody response against peptide epitopes above threshold from SARS-CoV-2 proteome during COVID-19 disease progression in three patients. (**a–f**) Evolution of positive IgA and IgG responses against SARS-CoV-2 peptides at different time points after onset of disease in patients #1, #2, and #3. Data were generated with peptide microarrays containing the whole SARS-CoV-2 proteome as 4883 overlapping peptides. Fluorescence intensities were transformed with the inverse hyperbolic sine (asinh function). Threshold selection shown as dashed horizontal lines: 99.9th percentile of IgA/IgG signals in healthy control sample #3 d-180, IgA: 6.54 transformed arbitrary fluorescence units, and IgG: 7.68 transformed arbitrary fluorescence units. For full array data, see Supplementary Materials B.

Patient #1 showed IgA responses at d6 (Figure 2a and Supplementary Materials A Table S2), solely targeting NSP2. The response was still limited four days later (d10), but then developed into a broad response at d22, targeting the nonstructural (NSP), the spike (S), membrane (M), ORF8, and nucleocapsid (N) proteins. The number of identified IgA-specific epitopes found in the different proteins increased over time in patient #1, with a particularly strong response to NSP3 and NSP12, while three signals for NSP2 epitopes decreased considerably. Regarding IgG responses (Figure 2b), patient #1 developed antibodies targeting the S and M protein already at d6 and the number of detected epitopes increased until d22.

In comparison, patient #2 (mild case, Figure 2c) showed a stronger and more specific IgA response already at d3 against the S, E, N, and NS proteins, while the IgG response (Figure 2d) revealed binding to NSPs and S. Patient #3 showed a strong and early response in IgA against many NSPs and the S protein (Figure 2e) comparing d-180 and d4, while the IgG (Figure 2f) only showed an increase in binding to NSP3 and S.

We visualized the identified epitopes derived from all patients (nine patient samples vs. control sample #3 d-180) on the S, M, and N proteins (Figure 3 and Supplementary Materials A Table S3). The data showed generally more IgA than IgG or IgM epitopes, with most epitopes located in the S and N protein.

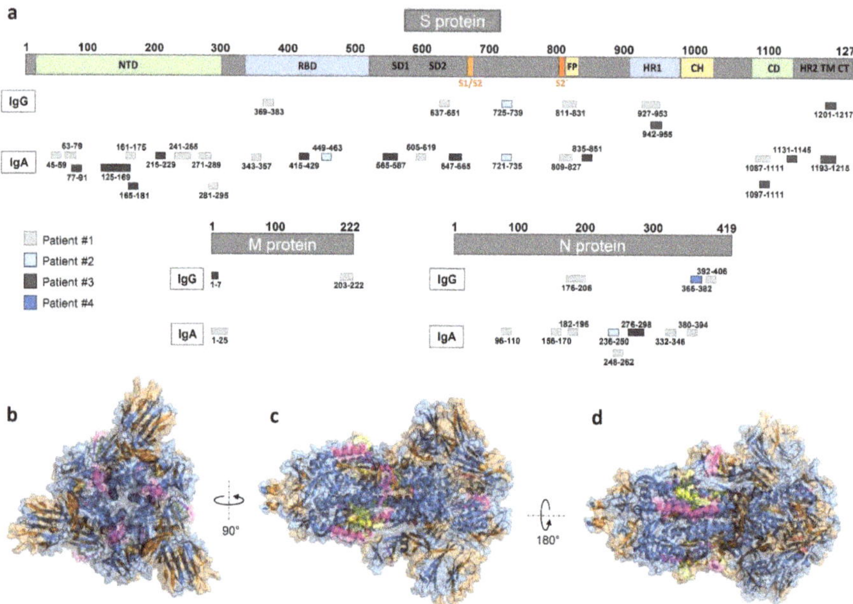

Figure 3. Mapping of peptide microarray reactive antibodies on the S, M, and N-proteins. (**a**) IgG and IgA epitopes derived from all four patients on the S (domains from [15]), M, and N proteins (see Supplementary Materials A Table S3). (**b–d**) A 3D view of the S GP structure (in blue) with the herein identified IgA (highlighted in orange), IgG (highlighted in magenta), and overlapping (highlighted in yellow) epitopes derived from all nine SARS-CoV-2 patient samples (generated with PyMOL from Protein Data Bank file 6vxx—cryo-EM structure of S GP [16]). (**b**) Top view, (**c**) side view, and (**d**) alternate side view of S GP trimer.

Finally, serum samples were analyzed with glycan microarrays, covering a diverse library of glycans (Figure 4, detailed information in Supplementary Materials C and Supplementary Materials A Figure S1). The glycans on the arrays cover several epitopes of the glycan shield of the SARS-CoV-2 surface [17]. Strong binding could be observed for the N-glycan core fragment ($Man_2GlcNAc_2$). Furthermore, α1-2-Man_3 showed increased binding in convalescent time points, which hints to binding of high-mannose (M7–M9) structures, which are reported to be part of the glycan shield of SARS-CoV-2 (especially spike N234) [17]. Similar to the observed trend with the peptide microarrays, patient #1 showed a strong increase in antibody binding at d22 towards the *N*-glycan core structures (strongest increase observed in IgM), while patient #2 had generally stronger and more constant signals, except for the binding to α1-2-Man$_3$. These results confirmed the general trends observed in both peptide and glycan microarray approaches, showing a strong increase of antibody responses in patient #1 at d22.

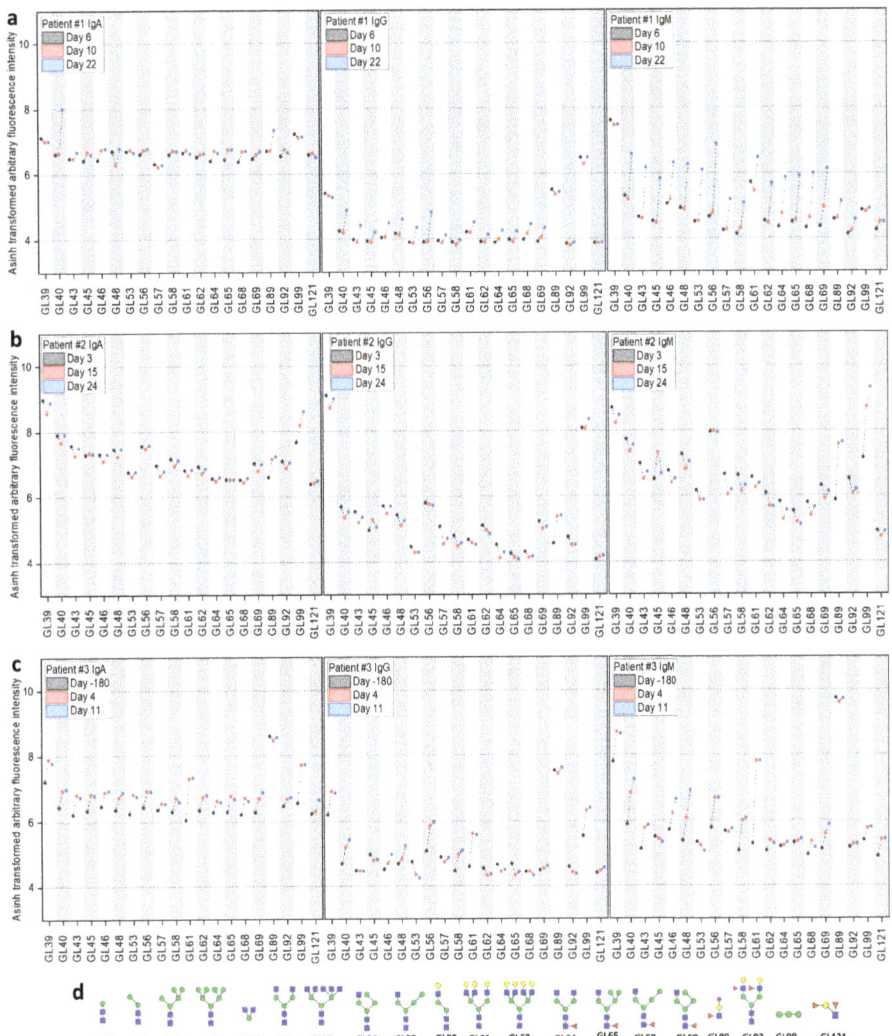

Figure 4. Longitudinal IgA, IgG, and IgM antibody response analyzed with glycan microarrays during COVID-19 disease progression in three patients. Evolution of reactive IgA, IgG, and IgM antibodies in patients #1–#3 (**a–c**) against a selection of glycans (**d**) on the microarrays at different time points (see Supplementary Materials C). For improved visualization, fluorescence intensities were transformed with the inverse hyperbolic sine (asinh function).

4. Discussion

To better understand the development of antibodies in SARS-CoV-2 infection and their consequence on the course of disease, we evaluated three assays that monitor the kinetics of B-cell epitope development in relation to clinical features. Three patients were sampled longitudinally. Two patients experienced a mild disease course (#2 and #3), another a moderate course (#1). We compared the longitudinal antibody response data using spike glycoprotein ELISA, peptide arrays of the whole SARS-CoV-2 proteome, and glycan arrays.

Serology testing for COVID-19 using ELISA is attractive because of the relatively short time to diagnosis and the ability to test for an active immune response against the virus. Comparing the three different approaches, the ELISA gives rather robust signals in the convalescent phase, while failing for early antibody detection. It is possible that the ELISA detects antibodies that either bind conformational/discontinuous epitopes or to glycopeptides. The assay gave similar positive results for both patients #1 and #2 in the convalescent phase (days 22 and 24). Comparing the data of patient #2 with mild symptoms we found that the peptide array shows a decline of binding to linear peptides over time, while the ELISA points into the opposite direction. In contrast, patient #1, with moderate symptoms, shows much stronger signals to the linear peptides on the array over time, which corresponds to the ELISA data. However, ELISA was inconclusive for patient #3, resulting in generally positive IgA and generally negative (or intermediate) IgG results for all time points, including d-180 prior infection and d32 (only analyzed by ELISA).

Recently, early antibody responses have been reported by ELISA [18], where seroconversion was found on day 7 after onset of symptoms in 50% of analyzed individuals. Another study underlined the early responses of IgA, IgM, and IgG following SARS-CoV-2 infection [19]. The authors reported a median duration of IgM and IgA antibody detection of five days and the detection of IgG 14 days after disease onset. Furthermore, Okba et al. [8] analyzed IgA and IgG responses in two mild and one moderate case using an in-house S1-ELISA. They observed an increase in the IgA response over time in a moderate case. An early or increased IgA response on arrays, as seen in our patients #2 and #3, was not observed with ELISA, possibly due to differences in the patients (sample collection dates) or assay performance. Key differences in these assays are the limitation of only using S1-proteins for the ELISA (vs. whole proteome on the microarray) and a higher sensitivity of peptide arrays towards linear epitopes. With the peptide arrays, cross-reactions to previous infections (e.g., with other coronaviruses) may become visible.

In contrast to ELISA, arrays are more time- and cost-intensive but provide more information on the development of antibodies. We identified several spike protein epitopes that are bound by IgA antibodies. We identified spike-specific IgA epitopes in the receptor binding domain, AA343-357, AA415-429, and AA449-463. The latter epitope is located in the receptor binding domain-angiotensin-converting enzyme II (RBD-ACE2)-complex and, therefore, may be the target of neutralizing antibodies [20]. In addition, we also confirm a part (AA369-383) of the SARS-CoV-2 and SARS-CoV cross-reactive IgG epitope (AA369-392) identified by Yuan et al., which is located in the receptor binding domain of spike [21]. Next, we observed IgA (AA809-827) and IgG (AA811-831) antibody binding, corresponding to the S2 cleavage site and fusion peptide. These have been described as distinctive epitopes in COVID-19 patients with neutralizing potential [20,22]. Furthermore, we could identify reactive peptides, especially in the N protein, as well as NSP3 and NSP12. Data from a partial proteome array approach was reported [23], which confirms strong binding to the N protein, although they did not cover NSP3 and NSP12. In contrast to NSP-binding antibodies, which could be cross-reactive from other viral infections, antibodies binding structural proteins like the S and N proteins, could be more distinctive for a SARS-CoV-2 infection [22]. It will be of interest to determine the longevity of these antibody responses and its impact on neutralization [24].

With the peptide arrays, we detected an early IgA response in the mild cases (patients #2 and #3). Respiratory viruses can induce efficient IgA responses in secretions as well as in sera. It was proposed that an early IgA response is predominant in COVID-19 and is more effective in SARS-CoV-2 neutralization than IgG [25]. IgA antibodies might be valuable diagnostic markers for early SARS-CoV-2 identification especially in mild-symptom patients. Due to high sensitivity and specificity, arrays may be relevant as diagnostics for the detection of these early antibody responses. Patient #2 potentially benefited from her early IgA response, which led to a mild course of the disease.

Employing glycan arrays, we identified several glycans that correspond to small fragments of the *N*-glycan core (e.g., $Man_2GlcNAc_2$). In addition, we observed an increase in binding to $\alpha1$-2-Man_3 (GL99 on the array) in patients #2 and #3. This fragment is part of the antennae of high-mannose (M7–M9) *N*-glycans, present on the spike protein (e.g., N122, N234, N343, and possibly others) [17,26]. A promising, but technically overly challenging approach, would be to screen glycopeptides with native glycan structures. Casalino et al. highlighted the modulating role of the spike protein N-glycan sites N165 and N234 for the conformation of the RBD [27]. Furthermore, a neutralizing antibody has been identified that binds a larger glycopeptide epitope of the SARS-CoV-2 spike protein [28]. Interestingly, we observe many spike-related peptide epitopes on the array, which would carry an *N*-glycosylation on the native virus (e.g., patient #1 in spike: AA63-79, AA271-289, AA343-357, AA605-619, and AA1087-1111). The glycan arrays generally show a similar trend as the peptide array results: the antibody response increases in patient #1 over time, whereas it stays constant or decreases over the course of the infection in patients #2 and #3, except for $\alpha1$-2-Man_3. Since many microorganisms express $\alpha1$-2-Man_3 on their surface, the SARS-CoV-2 infection might have caused a boost of a pre-existing immune response towards this epitope. Yet, data has to be evaluated in a broader context, since signals to glycans may be part of an unrelated cross-reaction or response to a larger glycopeptide epitope and multivalency can strongly influence the results.

We screened longitudinal serum samples of COVID-19 patients with different methods to get insights into their antibody responses and compared our data with findings of recent literature. A clear limitation of our study is the number of subjects, but still we were able to observe trends for the development of antibodies early after SARS-CoV-2 infection. Since all samples were collected from the same cluster of infection, which were the first detected SARS-CoV-2 infections in Hamburg, Germany, a clear chain of infection could be assured and samples could be collected repeatedly. This, and the limited access to arrays (especially glycan arrays), restricted the cohort size.

Our study emphasizes the importance of microarrays for early diagnostics and understanding of antibody development following SARS-CoV-2 infection. Arrays are able to reveal heterogeneous antibody responses in patients with different severity of symptoms. With a high assay sensitivity, antibody development in patients can be tracked during the course of disease and also early after infection. A general limitation of arrays is the use of exclusively linear peptides, which cannot identify antibodies that bind conformational or discontinuous epitopes. We exclusively considered the initially published Wuhan strain without mutations, but can quickly incorporate these mutations into the assay, since the array production method is rapid and flexible [29].

With the limitations listed above, our study contributes to the understanding of differences in the course of disease. There is still limited understanding of the immune correlates of protection. Collectively, we present an analysis of longitudinal antibody response in serum samples, comparing the degree of disease severity with three different approaches.

Supplementary Materials: The following are available online at https://www.mdpi.com/article/10.3390/pathogens10040438/s1. Supplementary A: Additional Information; Supplementary B: Peptide Microarray Data; Supplementary C: Glycan Microarray Data.

Author Contributions: J.H. performed all microarray related experiments and analyses. C.D. performed the immunological analyses. R.K., R.S., T.K., A.F., S.S. (Stefan Schmiedel), D.S., M.L.L. and the ID-UKE COVID-19 study group performed clinical monitoring and immunological analyses. F.F.L. supervised the microarray experiments, J.H. and F.F.L. performed the data analysis. N.-C.R., S.S. (Sonia Serna) and P.H.S. provided the glycan microarrays, B.M.S.S. and K.S. helped with their analysis. D.S. performed ELISA analyses. C.D., M.M.A. and F.F.L. supervised the project. C.D., J.H.

and F.F.L. wrote the manuscript. All authors helped in developing and revising the manuscript. All authors have read and agreed to the published version of the manuscript.

Funding: This research was supported by the German Federal Ministry of Education and Research [BMBF, Grant number 13XP5050A], the MPG-FhG cooperation [Glyco3Display], the Max Planck Society, the German Center for Infection Research [DZIF TTU01921], the Agencia Estatal de Investigación (Spain) [CTQ2017-90039-R], and the Maria de Maeztu Units of Excellence Program [MDM-2017-0720].

Institutional Review Board Statement: The protocol was approved by the Ethics Committee of the Hamburg Medical Association, Germany (PV7298).

Informed Consent Statement: Donors provided written informed consent.

Data Availability Statement: The data that support the findings are available online as supporting information or upon reasonable request from the corresponding authors.

Acknowledgments: We thank all participants for their involvement in and commitment to this study. We also thank Christina Lehrer for providing input and Anneke Novak-Funk and Elaine Hussey for critically reviewing the manuscript. In addition, we thank Ralf Bischoff, Gregor Jainta, and the company PEPperPRINT for their support, as well as Sabrina Kress and Jennifer Wigger for technical support. ID-UKE-COVID-19 Study Group: Marylyn M. Addo, Etienne Bartels, Thomas T. Brehm, Christine Dahlke, Anahita Fathi, Monika Friedrich, Svenja Hardtke, Till Koch, Ansgar W. Lohse, My L. Ly, Veronika Schlicker, Stefan Schmiedel, L. Marie Weskamm, and Julian Schulze zur Wiesch (University Medical Center Hamburg-Eppendorf).

Conflicts of Interest: The authors do not have any conflicts of interest to declare. The funding organizations did not play any role in the study design; data collection, analyses, or interpretation; the writing of the manuscript; or in the decision to publish the results.

References

1. Zhou, P.; Yang, X.-L.; Wang, X.-G.; Hu, B.; Zhang, L.; Zhang, W.; Si, H.-R.; Zhu, Y.; Li, B.; Huang, C.-L.; et al. A pneumonia outbreak associated with a new coronavirus of probable bat origin. *Nature* **2020**, *579*, 270–273. [CrossRef]
2. Peiris, J.; Lai, S.; Poon, L.; Guan, Y.; Yam, L.; Lim, W.; Nicholls, J.; Yee, W.; Yan, W.; Cheung, M.; et al. Coronavirus as a possible cause of severe acute respiratory syndrome. *Lancet* **2003**, *361*, 1319–1325. [CrossRef]
3. Ksiazek, T.G.; Erdman, D.; Goldsmith, C.S.; Zaki, S.R.; Peret, T.; Emery, S.; Tong, S.; Urbani, C.; Comer, J.A.; Lim, W.; et al. A Novel Coronavirus Associated with Severe Acute Respiratory Syndrome. *N. Engl. J. Med.* **2003**, *348*, 1953–1966. [CrossRef]
4. Drosten, C.; Günther, S.; Preiser, W.; Van Der Werf, S.; Brodt, H.R.; Becker, S.; Rabenau, H.; Panning, M.; Kolesnikova, L.; Fouchier, R.A.M.; et al. Identification of a Novel Coronavirus in Patients with Severe Acute Respiratory Syndrome. *N. Engl. J. Med.* **2003**, *348*, 1967–1976. [CrossRef]
5. Zaki, A.; Van Boheemen, S.; Bestebroer, T.; Osterhaus, A.; Fouchier, R. Isolation of a Novel Coronavirus from a Man with Pneumonia in Saudi Arabia. *N. Engl. J. Med.* **2012**, *367*, 1814–1820. [CrossRef]
6. Dong, E.; Du, H.; Gardner, L. An interactive web-based dashboard to track COVID-19 in real time. *Lancet Infect. Dis.* **2020**, *20*, 533–534. [CrossRef]
7. Guan, W.J.; Ni, Z.Y.; Hu, Y.; Liang, W.H.; Ou, C.Q.; He, J.X.; Liu, L.; Shan, H.; Lei, C.L.; Hui, D.S.C.; et al. Clinical Characteristics of Coronavirus Disease 2019 in China. *N. Engl. J. Med.* **2020**, *382*, 1708–1720. [CrossRef]
8. Okba, N.M.; Müller, M.A.; Li, W.; Wang, C.; GeurtsvanKessel, C.H.; Corman, V.M.; Lamers, M.M.; Sikkema, R.S.; De Bruin, E.; Chandler, F.D.; et al. Severe Acute Respiratory Syndrome Coronavirus 2—Specific Antibody Responses in Coronavirus Disease Patients. *Emerg. Infect. Dis.* **2020**, *26*, 1478–1488. [CrossRef] [PubMed]
9. Szymczak, L.C.; Kuo, H.-Y.; Mrksich, M. Peptide Arrays: Development and Application. *Anal. Chem.* **2018**, *90*, 266–282. [CrossRef]
10. Jaenisch, T.; Heiss, K.; Fischer, N.; Geiger, C.; Bischoff, F.R.; Moldenhauer, G.; Rychlewski, L.; Sié, A.; Coulibaly, B.; Seeberger, P.H.; et al. High-density Peptide Arrays Help to Identify Linear Immunogenic B-cell Epitopes in Individuals Naturally Exposed to Malaria Infection. *Mol. Cell. Proteom.* **2019**, *18*, 642–656. [CrossRef] [PubMed]
11. Geissner, A.; Seeberger, P.H. Glycan Arrays: From Basic Biochemical Research to Bioanalytical and Biomedical Applications. *Annu. Rev. Anal. Chem.* **2016**, *9*, 223–247. [CrossRef]
12. Muthana, S.M.; Gildersleeve, J.C. Glycan microarrays: Powerful tools for biomarker discovery. *Cancer Biomark.* **2014**, *14*, 29–41. [CrossRef]
13. Echeverria, B.; Serna, S.; Achilli, S.; Vivès, C.; Pham, J.; Thépaut, M.; Hokke, C.H.; Fieschi, F.; Reichardt, N.-C. Chemoenzymatic Synthesis of N-glycan Positional Isomers and Evidence for Branch Selective Binding by Monoclonal Antibodies and Human C-type Lectin Receptors. *ACS Chem. Biol.* **2018**, *13*, 2269–2279. [CrossRef]
14. Pfefferle, S.; Günther, T.; Kobbe, R.; Czech-Sioli, M.; Nörz, D.; Santer, R.; Oh, J.; Kluge, S.; Oestereich, L.; Peldschus, K.; et al. Low and high infection dose transmission of SARS-CoV-2 in the first COVID-19 clusters in Northern Germany. *medRxiv* **2020**. [CrossRef]

15. Wrapp, D.; Wang, N.; Corbett, K.S.; Goldsmith, J.A.; Hsieh, C.-L.; Abiona, O.; Graham, B.S.; McLellan, J.S. Cryo-EM structure of the 2019-nCoV spike in the prefusion conformation. *Science* **2020**, *367*, 1260–1263. [CrossRef]

16. Walls, A.C.; Park, Y.J.; Tortorici, M.A.; Wall, A.; McGuire, A.T.; Veesler, D. Structure, function, and antigenicity of the SARS-CoV-2 spike glycoprotein. *Cell* **2020**, *181*, 281–292. [CrossRef] [PubMed]

17. Watanabe, Y.; Allen, J.D.; Wrapp, D.; McLellan, J.S.; Crispin, M. Site-specific glycan analysis of the SARS-CoV-2 spike. *Science* **2020**, *369*, 330–333. [CrossRef]

18. Wölfel, R.; Corman, V.M.; Guggemos, W.; Seilmaier, M.; Zange, S.; Müller, M.A.; Niemeyer, D.; Jones, T.C.; Vollmar, P.; Rothe, C.; et al. Virological assessment of hospitalized patients with COVID-2019. *Nat. Cell Biol.* **2020**, *581*, 465–469. [CrossRef]

19. Guo, L.; Ren, L.; Yang, S.; Xiao, M.; Chang, D.; Yang, F.; Cruz, C.S.D.; Wang, Y.; Wu, C.; Xiao, Y.; et al. Profiling Early Humoral Response to Diagnose Novel Coronavirus Disease (COVID-19). *Clin. Infect. Dis.* **2020**, *71*, 778–785. [CrossRef]

20. Wang, H.; Wu, X.; Zhang, X.; Hou, X.; Liang, T.; Wang, D.; Teng, F.; Dai, J.; Duan, H.; Guo, S.; et al. SARS-CoV-2 Proteome Microarray for Mapping COVID-19 Antibody Interactions at Amino Acid Resolution. *ACS Cent. Sci.* **2020**, *6*, 2238–2249. [CrossRef]

21. Yuan, M.; Wu, N.C.; Zhu, X.; Lee, C.C.D.; So, R.T.Y.; Lv, H.; Mok, C.K.P.; Wilson, I.A. A highly conserved cryptic epitope in the receptor binding domains of SARS-CoV-2 and SARS-CoV. *Science* **2020**, *368*, 630–633. [CrossRef] [PubMed]

22. Shrock, E.; Fujimura, E.; Kula, T.; Timms, R.T.; Lee, I.-H.; Leng, Y.; Robinson, M.L.; Sie, B.M.; Li, M.Z.; Chen, Y.; et al. Viral epitope profiling of COVID-19 patients reveals cross-reactivity and correlates of severity. *Science* **2020**, *370*, eabd4250. [CrossRef]

23. Jiang, H.W.; Li, Y.; Zhang, H.; Wang, W.; Yang, X.; Qi, H.; Li, H.; Men, D.; Zhou, J.; Tao, S.C. SARS-CoV-2 proteome microarray for global profiling of COVID-19 specific IgG and IgM responses. *Nat. Commun.* **2020**, *11*, 3581. [CrossRef]

24. Seow, J.; Graham, C.; Merrick, B.; Acors, S.; Steel, K.J.A.; Hemmings, O.; O'Bryne, A.; Kouphou, N.; Pickering, S.; Galao, R.P.; et al. Longitudinal evaluation and decline of antibody responses in SARS-CoV-2 infection. *medRxiv* **2020**. [CrossRef]

25. Sterlin, D.; Mathian, A.; Miyara, M.; Mohr, A.; Anna, F.; Claër, L.; Quentric, P.; Fadlallah, J.; Devilliers, H.; Ghillani, P.; et al. IgA dominates the early neutralizing antibody response to SARS-CoV-2. *Sci. Transl. Med.* **2021**, *13*, eabd2223. [CrossRef]

26. Shajahan, A.; Supekar, N.T.; Gleinich, A.S.; Azadi, P. Deducing the N-and O-glycosylation profile of the spike protein of novel coronavirus SARS-CoV-2. *Glycobiology* **2020**, *30*, 981–988. [CrossRef]

27. Casalino, L.; Gaieb, Z.; Goldsmith, J.A.; Hjorth, C.K.; Dommer, A.C.; Harbison, A.M.; Fogarty, C.A.; Barros, E.P.; Taylor, B.C.; McLellan, J.S.; et al. Beyond Shielding: The Roles of Glycans in the SARS-CoV-2 Spike Protein. *ACS Cent. Sci.* **2020**, *6*, 1722–1734. [CrossRef]

28. Pinto, D.; Park, Y.J.; Beltramello, M.; Walls, A.C.; Tortorici, M.A.; Bianchi, S.; Jaconi, S.; Culap, K.; Zatta, F.; de Marco, A.; et al. Cross-neutralization of SARS-CoV-2 by a human monoclonal SARS-CoV antibody. *Nature* **2020**, *583*, 290–295. [CrossRef]

29. Stadler, V.; Felgenhauer, T.; Beyer, M.; Fernandez, S.; Leibe, K.; Güttler, S.; Gröning, M.; König, K.; Torralba, G.; Hausmann, M.; et al. Combinatorial Synthesis of Peptide Arrays with a Laser Printer. *Angew. Chem. Int. Ed.* **2008**, *47*, 7132–7135. [CrossRef] [PubMed]

pathogens

Review

Filovirus Neutralising Antibodies: Mechanisms of Action and Therapeutic Application

Alexander Hargreaves [1,2], Caolann Brady [1], Jack Mellors [1,3,4], Tom Tipton [1], Miles W. Carroll [1,3] and Stephanie Longet [1,*]

[1]   Nuffield Department of Medicine, Wellcome Centre for Human Genetics, University of Oxford, Oxford OX3 7BN, UK; Alexander.Hargreaves@well.ox.ac.uk (A.H.); Caolann.Brady@well.ox.ac.uk (C.B.); Jack.Mellors@phe.gov.uk (J.M.); Tom.Tipton@well.ox.ac.uk (T.T.); miles.carroll@ndm.ox.ac.uk (M.W.C.)
[2]   Faculty of Health and Medical Sciences, University of Surrey, Guildford GU2 7XH, UK
[3]   National Infection Service, Public Health England, Porton Down, Salisbury SP4 0JG, UK
[4]   Department of Infection Biology, Institute of Infection and Global Health, University of Liverpool, Liverpool L69 7ZX, UK
*   Correspondence: stephanie.longet@well.ox.ac.uk; Tel.: +44-18-6561-7892

Abstract: Filoviruses, especially Ebola virus, cause sporadic outbreaks of viral haemorrhagic fever with very high case fatality rates in Africa. The 2013–2016 Ebola epidemic in West Africa provided large survivor cohorts spurring a large number of human studies which showed that specific neutralising antibodies played a key role in protection following a natural Ebola virus infection, as part of the overall humoral response and in conjunction with the cellular adaptive response. This review will discuss the studies in survivors and animal models which described protective neutralising antibody response. Their mechanisms of action will be detailed. Furthermore, the importance of neutralising antibodies in antibody-based therapeutics and in vaccine-induced responses will be explained, as well as the strategies to avoid immune escape from neutralising antibodies. Understanding the neutralising antibody response in the context of filoviruses is crucial to furthering our understanding of virus structure and function, in addition to improving current vaccines & antibody-based therapeutics.

Keywords: neutralising antibodies; filoviruses; ebolavirus; post-vaccination; post-infection; monoclonal antibodies; longitudinal antibody response

Citation: Hargreaves, A.; Brady, C.; Mellors, J.; Tipton, T.; Carroll, M.W.; Longet, S. Filovirus Neutralising Antibodies: Mechanisms of Action and Therapeutic Application. *Pathogens* **2021**, *10*, 1201. https://doi.org/10.3390/pathogens10091201

Academic Editors: Philipp A. Ilinykh and Kai Huang

Received: 20 July 2021
Accepted: 12 September 2021
Published: 16 September 2021

Publisher's Note: MDPI stays neutral with regard to jurisdictional claims in published maps and institutional affiliations.

1. Filoviridae Background

1.1. Filoviridae Phylogeny

The first filovirus genus to be identified was *Marburgvirus* in 1967 composed of one species *Marburg marburgvirus* with two viruses: Marburg virus (MARV) and the very closely related Ravn virus (RAVV). Both viruses cause Marburg virus disease (MVD), a highly lethal form of viral haemorrhagic fever, with the largest outbreak occurring between 2004 and 2005 in Angola, with 252 infected individuals and a case fatality rate (CFR) of 90% [1].

The second and most notorious genus of filovirus, *Ebolavirus*, contains six species each with one virus; *Zaire ebolavirus*, Ebola virus (EBOV); *Sudan ebolavirus*, Sudan virus (SUDV); *Bundibugyo ebolavirus*, Bundibugyo virus (BDBV); *Tai Forest ebolavirus*, Tai Forest virus (TAFV); *Reston ebolavirus*, Reston virus (RESTV) and *Bombali ebolavirus*, Bombali virus (BOMV). The first four of these six viruses are known to cause Ebola virus disease (EVD) in humans, with a CFR frequently reported between 40% and 90%. However, this is likely an overestimate as many EBOV infections may go unreported [2]. EBOV is predominantly responsible for the EVD outbreaks of the greatest magnitude, with the largest being the 2013–2016 West African outbreak [3,4].

Most recently, new filoviruses have been discovered, which have not yet been associated with outbreaks in humans. In 2011, a third genus *Cuevavirus* was discovered with a sole species *Lloviu cuevavirus* including one virus, Lloviu virus (LLOV) [5]. Whilst infectious LLOV still remains to be isolated, anti-LLOV antibodies have been detected in bats [6,7]. In 2019, a new filovirus species *Měnglà dianlovirus*, including Mengla virus (MLAV) was discovered in China as the sole species of a new genus *Dianlovirus* [8]. Still to date, neither BOMV, LLOV nor MLAV are known to cause viral haemorrhagic fever with its pathogenicity in humans still to be determined [8,9]. However, both BOMV and MLAV Glycoprotein (GP) pseudoviruses as well as LLOV virus-like particles (VLPs) demonstrate a broad tissue tropism in cell lines from different animals, replicate similarly to ebolaviruses and use the Niemann pick type C1 (NPC1) as an entry receptor indicating a potential for spillover events [3,8,10]. Two much more divergent species of filovirus *Huángjiāo thamnovirus* including Huángjiāo virus (HUJV) and *Xīlǎng striavirus* including Xīlǎng virus (XILV), which belong to the genera *Thamnovirus* and *Striavirus* respectively, have been also described. These filoviruses infect fish [11].

While non-EBOV filoviruses will be mentioned, this review will be more focused on EBOV due to the ongoing impact of this pathogen on world health and the recent developments in antibody-based therapeutics and EBOV vaccines.

1.2. Genome Organisation of Ebolavirus

With the exception of the more divergent *Thamnovirus* and *Striavirus* genera, all filoviruses encode seven structural proteins: Nucleoprotein (NP), Viral Protein (VP) VP35, VP40, GP, VP30, VP24 and L polymerase (L) as shown in Figure 1A.

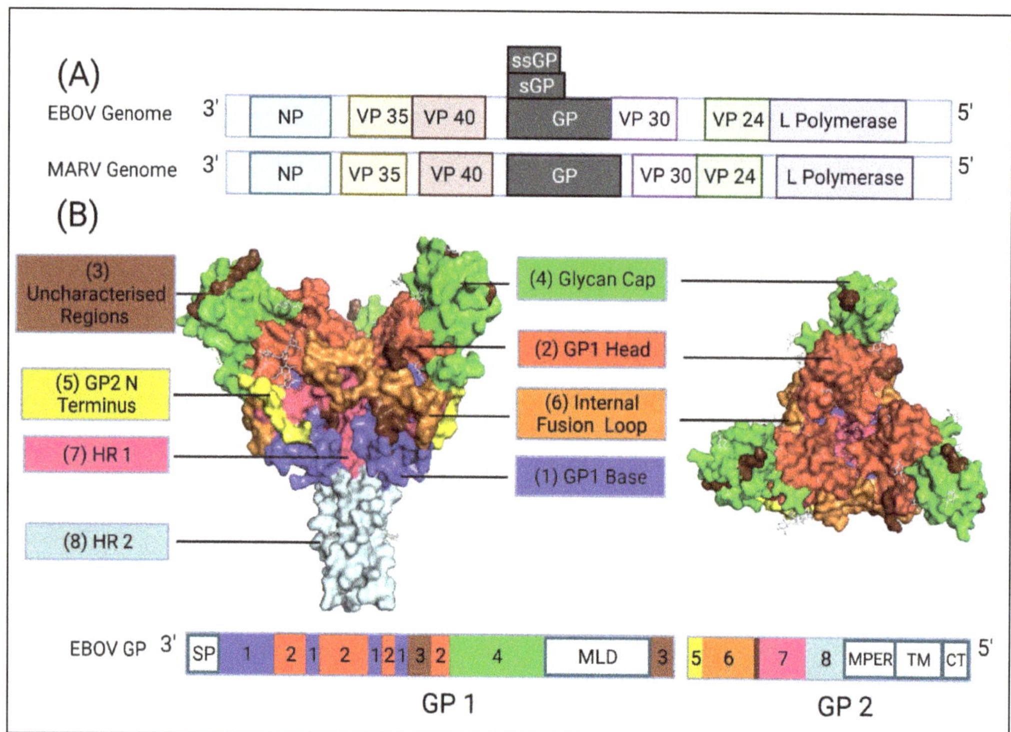

Figure 1. (**A**) A comparison of the EBOV and MARV genome structures highlighting key differences such as how EBOV

has multiple GP gene products and where there are differences in overlapping genes. (**B**) A diagram of the EBOV GP with domains highlighted based on the work of Lee et al. 2009 [12]. Domains of the EBOV GP as they appear on the gene: GP_1, SP (signal peptide), GP_1 base residues: (33–69, 95–104, 158–167 and 175–189), GP_1 Head: (70–94, 105–157, 168–175 and 214–226), Glycan Cap: (227–310), MLD Mucin-like-Domain, and in GP_2, N-terminus of GP_2, Internal Fusion Loop: (511–553), HR1 (554–598), HR2 (599–630), MPER Membrane proximal external region, TM transmembrane domain, CT Cytoplasmic tail. Regions highlighted in brown are uncharacterised as they could not be assigned to a domain due to differences with the protein structure used and the domains as listed by Lee et al. 2009 (32, 191–195, 210–211, 470–478) and the protein structure used from Zhao et al. [12,13]. Regions of the EBOV GP in white boxes with blue outlines are regions that could not be shown via X-ray diffraction and so do not appear on the GP structure in the diagram, in addition to residues (28–30, 196–209, 284–285, 294–300, 431–469, 632–669) [12,13]. Diagram was created with BioRender.com (©BioRender 2021, accessed in June 2021) and the protein structure was generated in The PyMOL Molecular Graphics System, Version 2.0 Schrödinger, Delano Scientific LLC, Berkeley, CA, USA using the PDB accession: 5JQ3.

The NP is the main component of the ribonucleoprotein or nucleocapsid, though VP30 and VP24 are also required for the stability of the nucleocapsid and together make the changes in NP required to incorporate the viral RNA. L polymerase is an RNA-dependent RNA polymerase which complexes with the polymerase co-factor VP35 and is responsible for transcribing the viral RNA, while the initiation of transcription is activated by VP30. L polymerase also functions in regulating and editing the viral RNA e.g., in the case of GP where three different gene products are produced [14]. VP40 is considered the matrix protein and is crucial for viral assembly and budding. It is also worth noting that many of the viral proteins, particularly VP24, VP30, VP35 and VP40 have functions linked to host pathology or immune evasion. For instance, VP24 and VP35 inhibit interferon (IFN) pathways (reviewed in Cantoni and Rossman, 2018) [15].

1.3. Cellular Entry of Ebola Virus

Cell entry is a critical stage in the lifecycle of any virus. GP exposed on the surface of the virion has been demonstrated to be essential for cell entry. In which case GP is the most important target for neutralising antibodies.

In EBOV, co- or post- translational editing (e.g., transcriptional slippage) of the GP transcript produce various GP products (reviewed in detail by Lee et al.) [12]. Similar post-translational modifications are not observed in MARV (Figure 1A). In EBOV, the pre-GP gene is transcribed as one protein but is cleaved by the furin protease into two subunits, GP_1 and GP_2, joined by a disulphide bridge to form a heterodimer [16]. Surface GP exists as a trimer of the GP heterodimers expressed on the surface of the virion and interacts with NPC1 for cell entry [17]. However, surface GP is not the main gene product. The primary product, which is expressed from unedited RNA transcripts, is a non-structural protein called secreted GP (sGP). Secreted GP is a dimer of GP_1 and a truncated GP_2 bound by disulphide bridges [18] which is hypothesised to act as a decoy antigen to sequester antibodies or cause antigenic subversion. Surface GP is only produced upon the addition of an adenosine residue to the GP transcript in 20% of cases. A deletion of one adenosine or addition of two adenosines in the transcript leads to the production of another non-structural protein, small secreted GP (ssGP) [12,19]. The role of ssGP is unclear in EBOV pathogenesis. Lastly, during EBOV infection, GP can be shed from infected cells in a soluble form due to cleavage by TACE metalloprotease. This soluble GP is named shed GP and was shown to sequester EBOV-specific neutralising antibodies directed towards GP [20].

The entire GP is required for cell entry. GP_1 is associated with cell attachment and receptor binding, whilst GP_2 has functions regarding fusion with the host cell membrane. Both subunits are therefore targets of neutralising antibodies. GP_1 contains four subdomains: the head, base, mucin-like domain (MLD) and glycan cap (GC). The GP_1 head, when arranged in its trimeric conformation forms a three lobed chalice containing the receptor binding domain (RBD). The GP_1 base clamps the internal fusion loop (IFL) and heptad repeat 1 (HR1) and heptad repeat 2 (HR2) from GP_2 for arrangement into its pre-fusion conformation (Figure 1B) [12]. There are also two heavily glycosylated regions on the GP_1, GC and the MLD [12]. The GC is a large chain of sugars, some of which are thought to act as attachment factors to host cells e.g., binding to C-type lectins. Both the MLD and GC are highly variable regions and can block neutralising antibodies binding to the GP, e.g., the MLD overhangs the GP to protect critical epitopes [21].

GP$_2$ is docked into the viral membrane by its transmembrane domain. GP$_2$ is also responsible for fusing the virus membrane and the host endosome membrane. It contains several domains that are key to this process such as HR1 and HR2 that change conformation to aid the IFL insertion into the host endosome membrane ultimately leading to fusion (Figure 1B) [12].

EBOV uses a broad range of binding factors to attach to a variety of host cell types before being internalised by the host cell into an endocytic compartment by receptor mediated endocytosis or macropinocytosis [12,22]. In the late endosome, cathepsins cleave the GP removing the MLD and GC and exposing the RBD in a stage called priming, allowing better access to the RBD for receptor binding [17,23]. The RBD then binds to the NPC1 cholesterol transporter [24]. The base of GP$_1$ is a clamp, that when released triggers the conformational changes required to expose the hydrophobic IFL which inserts into the endosomal membrane [12,25]. HR1 and HR2 pull on the IFL causing it to fuse with endosomal membrane, creating a pore for the release of the ribonucleoprotein into the cell cytoplasm [12,26].

1.4. EVD and Immunity

EVD is characterised as a systemic inflammatory response ("septic-shock-like-syndrome") with coagulation abnormalities and multi-organ failure, immunosuppression and lymphopenia [4]. This pathology is a result of GP-mediated activation of innate immune cells.

EBOV preferentially targets dendritic cells (DC), monocytes and macrophages but can also productively infect epithelial and endothelial cells, adrenal fibroblasts and hepatocytes [27]. GP-mediated activation of the TLR4 pathway in DCs and macrophages [28,29] results in the expression and secretion of inflammatory cytokines (e.g., IL-6, IL-1β, TNF) [30–33]. In macrophages, EBOV elicits a strong upregulation of IFN-I signalling [34], a crucial feature of the immune response to viral infection that provides the link between innate and adaptive immunity [35]. However, the role of Type I IFN in EBOV clearance and protection is not fully understood. Several studies demonstrated that IFN-I impacted EBOV replication in vitro and survival in animal models [36–40], while higher levels of IFN-α were associated with EVD fatal cases [41]. With that said, the picture is likely more complicated as EBOV has also developed some strategies to dampen innate immune responses [42].

Following the activation of the innate immune system, adaptive immune responses are induced. Early studies in individuals who succumbed to EVD described robust T cell activation, followed by a collapse in the T cell population [43,44]. More recent studies demonstrated the impact of T cell dynamics, kinetics and phenotype on EVD outcome [44–46]. It is known that T cells, particularly CD8$^+$ T cells, are important in viral infection. However, the role of CD4$^+$ and CD8$^+$ T cells in protection is still under discussion. For example, it was demonstrated that CD8$^+$ T cell deficient mice, but not CD4$^+$ T cell or B cell deficient mice, succumbed to a subcutaneous infection with a mouse-adapted EBOV strain suggesting a crucial role of CD8$^+$ T cells in protection [47]. Following the 2013–2016 West Africa epidemic, more studies analysed the activation of T cells in EVD survivors. We reported that the dominant CD8$^+$ polyfunctional T cell phenotype was IFNγ^+, TNF$^+$, IL-2$^-$ in EVD survivors in Guinea [48]. We also found that both CD4$^+$ and CD8$^+$ T cells contributed to specific T cell memory but with a differing cytokine profile. CD4$^+$ T cells produced IFNγ, TNFα and IL-2, whereas CD8$^+$ T cells only produced IFNγ and TNFα [49]. Sakabe et al. found that CD8$^+$ responses to the NP were immunodominant in survivors in Sierra Leone [50]. More details about T cell responses following a natural EBOV infection can be found in several reviews [46,51]. Regarding the humoral response, some studies have suggested that early development of IgM and IgG was associated with a positive outcome [52] while antibody deficiencies were reported in fatal cases [53]. Furthermore, monoclonal antibodies (mAbs) were isolated from EVD survivors, some of which were shown to be neutralising and protective in animal models [54–59]. However, it is still debated whether the antibody titres or the neutralisation activity of antibodies is the stronger forecaster of protection [60]. This review will discuss the role of neutralising antibody responses following a natural infection and in the context of therapeutics and vaccination.

2. Neutralising Antibodies Following a Natural Infection

2.1. Neutralising Antibodies against EBOV

Mechanisms of antibody activity include neutralisation and Fc receptor-mediated effector functions (e.g., antibody-dependent cellular cytotoxicity, which is the killing of a target cell coated with antibodies by an effector immune cell; antibody-dependent cellular phagocytosis, which is a mechanism of clearance of antibody-coated pathogens or tumour cells by macrophages and natural killer cells and antibody-dependent complement deposition, which is the deposition of a complement component on infected cells mediated by IgG or IgM).

In the context of EBOV infection, it is not fully clear which function is the most associated with protection [46]. However, the characterisation of neutralising antibodies isolated from survivors and their subsequent evaluation in animal challenge models helped to improve our insight into neutralising antibody responses. It is crucial to underline that a lot of studies that analysed antibody responses to EBOV focused on IgG, especially total IgG in serum. However, the neutralisation activity can also involve pentameric IgM and monomeric/dimeric IgA present in serum and particularly abundant in mucosae [61].

Some studies suggested that neutralising antibody levels were modest early post-infection in humans but increased over time. Luczkowiak et al. analysed the neutralising activity of the plasma of three 2013–2016 West Africa epidemic survivors using a lentiviral EBOV-GP pseudotyped infection assay. They found that neutralising antibody titres increased up to 9 months post-infection [62]. Another study in Western patients confirmed this. Williamson et al. analysed B cell responses in four acute *Ebolavirus*-infected patients that had been repatriated to the US during the 2013–2016 West Africa epidemic. They found that between 1 and 3 months post-recovery, there was a low frequency of EBOV-specific B cells encoding for antibodies that displayed low neutralising activity. However, one neutralising antibody isolated in this study led to protection in a mouse EBOV challenge model [63]. A high diversity of neutralising antibodies may be needed for an efficient neutralisation, which could explain the delay in the development of neutralising responses [46]. However, it was clearly observed in several studies that long-term survivors developed robust and sustained neutralising antibody responses mainly targeting GP. Antibodies able to neutralise a pseudotyped EBOV GP were detected in survivors from the 1976 Yambuku outbreak 40 years later [64]. In addition, mAb 114, a potent neutralising mAb against pseudotyped EBOV GP lentivirus particles was first isolated from a survivor of the 1995 Kikwit outbreak, 11 years post-recovery [54]. This mAb is now a component of the recently approved drug named Ebanga®. The presence of persistent neutralising antibodies was confirmed in a larger cross-sectional study. Halfmann et al. measured anti-EBOV-specific humoral responses in 214 survivors and 267 close contacts (including 56 healthcare workers) in Sierra Leone, 15–32 (median 28) months post-recovery [65]. The study revealed 97.7% of survivors had antibodies against at least one of the three antigens (GP, VP40, NP) with 85% harbouring antibodies against all three. Of the survivors with a detectable antibody response against GP, all but one had neutralising titres, typically in the range of 1:128 to 1:152, but some exceeded >1:2048 [65]. A longitudinal study performed by Thom et al. also confirmed the maintenance of total and neutralising antibody responses over the course of several years. Thom et al. performed a longitudinal study looking at 117 survivors and 66 contacts, sampling patients from 3 to 14 months post-discharge, with follow up collections 12 and 24 months later [48]. The study found similar findings to Halfmann et al., where at 3–14 months 96% of survivors developed IgG specific response and this correlated well with total antibody titres (r 0.85; $p < 0.0001$). Interestingly, 96% of survivors maintained high titres of neutralising antibody against live Ebola virus, with a mean of 1/174. This is ten-fold greater than the titres observed in patients one-month post-vaccination with an EBOV vaccine, however, it is not known what it means for the protection as correlates of protection are unclear [48]. A smaller longitudinal study investigating B cell responses carried out by Davis et al. followed the four 2013–2016 West Africa epidemic survivors repatriated to the US (previously mentioned) from discharge to 2 years post-infection. Davis et al. concluded that IgM levels typically declined after a few months, while IgG and IgA levels remained elevated [66]. This was hypothesised to be a result of antigen restimulation. However, opposing evidence by Thom et al. indicated no antigen stimulation of T cells was detected during their longitudinal study, suggesting there might be a different reason for the retention of high titres. This topic is still debated [48]. Davis et al. also characterised EBOV GP-specific monoclonal antibodies and they found that only a subset was capable of recognising cell-surface GP, a subset that contained neutralising antibodies [66]. Some research groups characterised the regions targeted by the neutralising antibodies. Bornholdt et al. analysed 349 monoclonal antibodies specific to GP isolated from a 2014 EBOV Zaire outbreak survivor. They found that 77% of these antibodies could neutralise live EBOV. After analysing the epitopes recognised by the neutralising antibodies, they reported that mAbs which targeted the GP stalk region proximal to the viral membrane were particularly effective at protecting mice against lethal EBOV challenge [59]. Recently, Khurana et al. described specific sites on GP mounting a neutralising antibody response in rabbits, which were protective in a lethal EBOV mouse model [67]. EBOV GP regions targeted by neutralising antibodies will be discussed in the Section 5.2.

Recently, Gunn et al. described the profile of humoral responses in EVD survivors from Sierra Leone. Interestingly, this study highlighted the development of both neutralising and polyfunctional IgG1 and IgA in survivors [68]. It is probable that there is a synergy between the neutralising activity and the innate immune effector functions via Fc receptor of antibodies. To our knowledge, it has not been shown in the context of EBOV, but it has been recently described in the context of SARS-CoV-2 [69].

The delay in neutralising antibody response early post-infection but long-term persistence of neutralising antibodies in survivors may suggest a role of neutralising antibodies in the clearance of the Ebola virus rather than in the early stages of infection resolution. The characterisation of GP epitopes inducing protective neutralising antibodies, as well as a better understanding of antibody polyfunctionality at systemic and mucosal levels could be very valuable and may help to improve the efficacy of current antibody-based therapeutics and vaccines.

2.2. Neutralising Antibodies against Non-EBOV Filoviruses

Aside from EBOV epidemics, SUDV, BDBV and MARV outbreaks make up the vast majority of other recorded human filovirus infections and so it is important to consider the analysis of immune responses to these viruses. This is especially true because historically their small outbreaks have often occurred in remote regions of Africa with limited health infrastructure which restricts the collection of patient samples and data. Consequently, our understanding of these diseases is less detailed and, despite being equally deadly, there are currently no licensed therapeutic options for non-EBOV filoviruses.

Sobarzo et al. analysed the neutralising antibody responses of survivors of the 2000–2001 Gulu [70] and 2012 Kibaale [71] outbreaks caused by SUDV. Serum samples were collected 12 years or 3 years post-recovery, respectively. Similarly to EBOV, a robust and persistent SUDV-specific antibody response, mainly targeting GP and NP, was observed in both cohorts. Five of five Kibaale survivors and five of six Gulu survivors displayed antibodies capable of neutralising the whole SUDV by plaque reduction neutralisation assay (PRNT) [70,71].

Marburg has been responsible for two outbreaks exceeding case numbers of 100. Stonier et al. performed a longitudinal study following PCR positive survivors of the 2012 MARV outbreak in Uganda sampling patients 9, 15, 21 and 27 months after the outbreak [72]. All survivors had an antibody response to MARV GP and NP and most had a response to VP30 and VP40. Whilst total antibody titres remained high in all patients throughout the study, at 9 months, only two of six survivors demonstrated, via PRNT, a neutralising antibody response between 1:20 and 1:40 which diminished by 21 months and disappeared by month 27 [72]. These results suggest a more rapid waning of neutralising antibodies following MARV infection compared to EBOV infection. Could this difference be linked to the presence of extra GP products with EBOV, which is not observed with MARV? The mechanisms behind these results are unclear.

Other researchers isolated monoclonal antibodies from peripheral blood B cells from survivors of the 2007 BDBV outbreak in Uganda. They found a large proportion of BDBV GP-specific monoclonal antibodies including some antibodies which strongly neutralised BDBV but also SUDV and EBOV using chimeric filoviruses [57].

Together these results show that robust neutralising antibody responses are also induced following an infection with non-EBOV filoviruses.

3. Antibody-Based Therapeutics and Vaccines
3.1. Monoclonal Antibodies

Therapeutic mAbs have been a field of great interest. Two EBOV specific mAb therapeutics, REGN-EB3 (Inmazeb™, Regeneron Pharmaceuticals, Tarrytown, NY, USA) and mAb114 (Ebanga™, Ridgeback Biotherapeutics, Miami, FL, USA) having recently been approved by the Food and Drug Administration (FDA, Silver Spring, MD, USA).

Monoclonal antibodies can be created in a laboratory or isolated from a specific B cell clone from survivors or vaccinated animals. Monoclonal antibody therapy is the passive transfer of immunity in the form of one, or a small number of, antibodies targeting a single epitope of a protein, usually GP [73]. This differs from polyclonal antibodies elicited from vaccination or infection where antibodies target a higher number of epitopes on the viral protein [73]. One of the advantages is that mAbs have an immediate and potent effect and can be used as a post-exposure therapeutic and can also be administered prophylactically with a relatively long half-life, e.g., mAb114 24.2 days [56,74].

The GP_1 head is home to the RBD and therefore an obvious target for mAbs including mAb 114. Mab114 is the only effective monotherapy and is licensed under the name Ebanga™ following impressive results from the Pamoja Tulinde Maisha (PALM) trial where it demonstrated a protective effect by reducing mortality to 35.1% [75,76]. Odesivimab is also part of the licensed Inmazeb™ cocktail but it is non-neutralising because it binds to the GP_1 head but a little further away from the RBD than mAb114 and consequently does not neutralise (Table 1) [77]. FVM04 is a broadly neutralising antibody that binds to the GP_1 head of the GP and can neutralise EBOV and SUDV in addition to providing good efficacy in mouse models (Table 1) [78].

The GP_1 base is potentially the most common target for neutralising antibodies. It is the target of KZ52, which is a research standard for neutralisation, even though it was shown not to be protective in NHP trials. KZ52 binds to an epitope on the GP_1 base that at least partially overlaps with several other potential therapeutics e.g., 2G4 and 4G7 [79]. 2G4 and 4G7 are both neutralising antibodies from the ZMapp cocktail which share overlapping epitopes on the GP_1 base (Table 1) [80]. Maftivimab neutralising antibody of the Inmazeb™ cocktail also binds to the GP_1 base. While it does not share an epitope with the aforementioned mAbs of the Inmazeb™ cocktail, it is thought to have the same mechanism of neutralisation (Table 1) [77].

The GC is a highly variable region that is usually thought of as non-neutralising, but it was found to be a target for several therapeutic mAbs. Saphire et al. via the Viral Haemorrhagic Fever Consortium demonstrated that antibodies targeting GC were mostly non-neutralising but a small subset of neutralising antibodies were detected [81]. The GC is also a target for Atoltivimab which is a neutralising antibody in the licensed Inmazeb™ cocktail and able to activate effector functions (Table 1) [77,81]. Furthermore, the GC contains the epitope for the broadly neutralising EBOV-548. This mAb binds to a conserved GC epitope in EBOV and BDBV. It can neutralise both viruses and effectively activates effector functions [82]. EBOV-520, which binds to the GP base region, strongly demonstrates co-operative binding. It binds to its epitope with $5\times$ greater affinity once EBOV-548 is bound to the GC. Together, EBOV-520 and EBOV-548 can fully protect NHPs from mortality, though EBOV-520 alone has also demonstrated partial protection in animal models [82–84]. Finally, 13C6 is a non-neutralising antibody from the ZMapp cocktail that binds to the GC but is removed during cathepsin cleavage. It also effectively activates Fc functions [80].

ADI-15878 is a neutralising pan-ebolavirus mAb which binds to conserved regions on the HR1 and IFL domains of EBOV, SUDV and BDBV [85]. When administered alone, survival rates post-challenge in guinea pigs of 33–50% were observed. However, when paired with ADI-23774 in a cocktail called MBP134 at a high dose, it provided 100% protection [86]. CA45 is another pan-ebolavirus mAb with neutralising activity against EBOV, SUDV, BDBV and RESTV that binds to the IFL and GP_1. This mAb was shown to be protective in animal models, both individually and in combination with FVM04 in NHP trials (Table 1) [87]. CA45 and FVM04 can also be given in a cocktail with MR191 an anti-MARV and RAVV mAb, that provides 100% protection against NHPs challenged with MARV when tested alone and within the cocktail [87,88]. MR191 binds to residues in the RBD that are essential for receptor binding and are hence highly conserved among related filoviruses. MR191 however, does not neutralise EBOV because the GC obscures the epitope whereas MARV has a disordered and flexible GC that gives the mAb access to this epitope [89,90].

The PALM trial is the only large clinical trial to date evaluating anti-EBOV experimental therapeutics in a field setting during the 2018 outbreak in the Congo, with patients presenting during various stages of disease following positive RT-PCR. It compared ZMapp, mAb114, REGN-EB3 and Remdesivir in 681 participants in a 1:1:1:1 ratio. One striking result was that ZMapp and Remdesivir were found to have no statistically significant effect on reducing EBOV mortality which stood around 50% [76]. The study did find however, that REGN-EB3 (Inmazeb™) and mAb114 (Ebanga™) had a protective therapeutic effect, reducing mortality to 33.5% and 35.1%, respectively. The effectiveness of the treatments was especially high among patients who received treatment early, with an estimated 11% increase in the risk of mortality with each day the treatment was delayed. Furthermore, patients who were classified to have a high viral titre also had a much higher mortality of 67% even with mAb114 or REGN-EB3 [76].

Table 1. A table describing the characteristics and epitopes of several key mAbs, with a figure integrated highlighting the residues on the EBOV GP that make up the epitopes of each of the mAbs in the table. The critical/known residues are in red, and the putative residues highlighted in yellow. The protein structure was generated in The PyMOL Molecular Graphics System, Version 2.0 Schrödinger, Delano Scientific LLC, Berkeley, CA, USA using the PDB accession: 5JQ3.

Antibody (Cocktail)	Epitope	Brief Description—How It Was Discovered and Where It Targets	Neutralising
KZ52—WHO Research Standard Critical residues—red: 511, 550, 552, 552, 556 Putative residues—yellow: 24, 40, 43, 507-508, 513-514, 549, 551		A potently neutralising antibody isolated from a survivor of the 1995 Kikwit outbreak binding to the GP_1/GP_2 interface to prevent insertion of the fusion loop into the membrane [80].	✓

Table 1. *Cont.*

Antibody (Cocktail)	Epitope	Brief Description—How It Was Discovered and Where It Targets	Neutralising
13C6 (ZMapp) Critical residues—red: 270, 272		A non-neutralising humanised mouse antibody that binds to the GP_1 head and GC regions and can activate effector functions [80].	X
2G4 (ZMapp) Critical residues—red: 511, 550, 553, 556		A neutralising humanised mouse antibody that binds to the GP_1/GP_2 interface to prevent insertion of the fusion loop into the endosome membrane [80].	✓
4G7 (ZMapp) Critical residues—red: 511, 552, 556		A neutralising humanised mouse antibody that binds to the GP_1/GP_2 interface to prevent insertion of the fusion loop into the membrane [80].	✓
Ansuvimab-zykl/mAb114 (Ebanga™) Known residues—red: 111-119		Isolated from a survivor of the 1995 Kikwit outbreak that targets the RBD blocking access to NCP1 [74,91].	✓
Atoltivimab/REGN3470 (Inmazeb™) Putative residues—yellow: 236-244, 264-297 and 298-308		Developed by the VelcoImmune platform in which genetically engineered mice express fully human antibodies. Atoltivimab is a neutralising antibody that binds to GC, it is also able to activate effector functions [92].	✓
Odesivimab/REGN3471 (Inmazeb™) Putative residues—yellow: 114 to 122, 139 to 151, 236 to 244, and 265 to 287		A non-neutralising VelcoImmune antibody that binds to the GP_1 head but further from the residues involved in receptor binding compared to mAb114. It is very capable of activating effector functions [92].	X

Table 1. *Cont.*

Antibody (Cocktail)	Epitope	Brief Description—How It Was Discovered and Where It Targets	Neutralising
Maftivimab/REGN3479 (Inmazeb™) Putative residues—yellow: 531 to 545		A neutralising VelcoImmune antibody that binds to the GP_1/GP_2 interface to prevent insertion of the IFL into the endosome membrane [92].	✓
CA45—Pan-ebolavirus Critical residues—red: 64, 517, 546, 550 Putative residues—yellow: 38, 40-41, 66, 68, 101-104, 184, 186-187, 211-213, 513-516, 518-519, 544-545, 547-549, 551-552, 554, 558		Isolated from challenged NHPs, CA45 binds to the IFL and GP_1/GP_2 interface preventing insertion of the IFL into the endosome membrane and hence neutralises EBOV, SUDV, BDBV and RESTV. CA45 provides protection in mice, guinea pigs and ferrets. Also gives 100% protection to NHPs when given in a cocktail with FVM04 and MR191 [87,93,94].	✓
FVM04—Pan-ebolavirus Known residues—red: 115, 117-118		Targets the RBD blocking interactions with NPC1. It neutralises EBOV, SUDV and BDBV, though only protective from EBOV and SUDV challenge in mouse and Guinea pig model. Can be given in cocktail with CA45 and MR191 [78,87].	✓

3.2. Convalescent Plasma

Convalescent plasma therapy is the passive transfer of immunity from a convalescent donor to a patient with an acute infection [95]. The transfer of convalescent blood product to a patient is an old therapy which was already used to treat various infections in humans and animal models at the end of 1800's [96]. The theoretical basis for its use against EBOV infection is that EVD survivors are thought to be protected from re-infection through humoral immunity and so the transfer of this protective sera containing EBOV-specific antibodies could have therapeutic benefits for patients. EBOV outbreaks take place in low-income countries where expensive drugs are not always affordable and so convalescent plasma may provide a more cost-effective treatment solution. However, it comes with its complications as donor blood must be tested for a wide variety of contaminants including HIV, malaria and hepatitis A which are known to have high prevalence in many regions that suffer EBOV outbreaks [95]. Convalescent plasma therapies were tested to reduce the risks of transfusion-transmitted infections and their impacts on anti-EBOV antibodies were assessed. For example, Amotosalen/UVA pathogen reduction technology was tested to treat EVD convalescent plasma. Plasma were analysed by two types of ELISA and three neutralisation assays and it was found that anti-EBOV titres remained relatively unchanged following the treatment [97].

During the 2013–2016 West Africa epidemic, the Ebola Tx trial in collaboration with The Conakry Ebola Survivors Association organised a large plasma collection programme in Conakry in Guinea between November 2014 and July 2015 [98]. This consortium evaluated the efficacy of convalescent plasma in comparison with standardised supportive care in EVD patients in a phase 2/3 open-label non-randomised trial setting. Even though no serious adverse effects were observed with the use of the convalescent plasma, the therapy was found to have no significant effect on mortality compared to the control group (31% of patients treated with convalescent plasma died 3–16 days post-diagnosis, compared to the 38% in the control group) [99]. These results may be due to the varying levels of EBOV-specific neutralising antibody in the plasma. At that time, they did not have any test to measure EBOV-specific antibody titres and neutralising antibodies in the field. Other clinical trials evaluated some methods to screen the antibody responses in donors' plasma in order to potentially select the plasma with high neutralising antibody titres. Brown et al. measured anti-EBOV antibodies in EVD survivors in Liberia in 2014–2015 using two ELISA and two neutralisation assays (microneutralisation, PRNT) [100]. They found that the four assays were concordant to measure donor antibody titres. However, 15 of 100 donors, including seven with a confirmed EBOV PCR positive result, did not have any detectable EBOV-specific antibodies. This trial found that viral load was reduced in EVD patients who received the convalescent plasma containing higher antibody levels, but not in patients who received the therapy with lower antibody levels [100]. Tedder et al. performed a similar evaluation in Sierra Leone using an IgG capture competitive double-antigen bridging enzyme immunoassay and a pseudotyped virus assay [101]. Both studies demonstrate the benefit of screening donors' plasma for neutralising antibody titres even though neutralising antibodies are yet to be validated as correlates of protection.

3.3. Vaccine-Induced Neutralising Responses

3.3.1. EBOV Vaccines

The role of antibody responses in protection following vaccination with a replication competent recombinant VSV virus encoding for Zaire EBOV Kikwit 1995 GP (rVSV-ZEBOV) was firstly analysed in preclinical studies. Marzi et al. vaccinated five NHP groups 28, 21, 14, 7 and 3 days before challenge [102]. Surprisingly, an induction of EBOV-GP specific IgG responses was already reported 3–7 days post-vaccination in NHPs. The research group found a partial protection of animals vaccinated 3 days before the challenge and a full protection of animals immunised $\geq$7 days before the challenge. This indicates that rVSV-ZEBOV may elicit very rapid humoral response and protection [102]. Another pre-clinical study reported that antibodies were sufficient at protecting mice from infection following immunisation with rVSV-ZEBOV. Meanwhile the depletion of CD8$^+$ T cells did not compromise protection [103]. Similar results were observed by Marzi et al. in NHPs. CD8$^+$-depletion did not impact on survival of rVSV-vaccinated animals following a challenge [104]. Wong et al. also analysed the role of humoral response following rVSV-ZEBOV vaccination in NHPs and found significantly higher total IgG titres in the serum of animals which survived post-challenge compared to the non-survivors [105]. The same team also confirmed long-term protection by challenging vaccinated animals 6–12 months post-immunisation with rVSV-ZEBOV. They observed that the levels of EBOV GP-specific IgG antibody, measured immediately before challenge, correlated with protection, whereas neutralising antibody were not always a reliable measure of protection in their animal model [106]. Another research group determined whether rVSV-ZEBOV vaccine could be used as a post-exposure treatment in Rhesus macaques. They found that four of eight macaques were protected if treated up to 30 min following a lethal infection. While the differences in cellular responses where minimal between the animals that survived and those that succumbed to the virus, there was a significant difference in neutralising responses. Indeed, neutralising antibodies were detected on days 14–36 post-challenge in animals that survived the infection, while the humoral response was not detected in animals that succumbed to the infection, suggesting a critical role for the humoral response [107]. Another study showed that NHPs vaccinated with rVSV-ZEBOV were also protected from EBOV aerosol challenge. Interestingly, upon measurement of circulating rVSV-ZEBOV specific-IgG responses post-vaccination and post-challenge, antibody responses were found to increase post-challenge. While the neutralising capacity of the antibodies was not analysed, they found there was no evidence of IFNγ or TNFα production in CD4$^+$ or CD8$^+$ before or after the challenge, which further suggests a major role of humoral response in protection [108]. Qiu et al., compared the protective immune responses in NHPs immunised with rVSV-ZEBOV by intramuscular (IM), intranasal (IN) or oral route (OR). They observed that IgG, IgA and IgM antibody responses were detected in the serum of all vaccinated animals independent of the route of vaccine administration. Post-challenge, IgG and IgA titres increased, while IgM titres did not exceed the levels observed post-vaccination. Globally, IgM titres ranked IN $\cong$ OR > IM, with an IN titre 2.4 times higher than IM, while IgA and IgG responses ranked IN > OR > IM (IN 6.8-fold > IM) and IN > OR $\cong$ IM (IN 9.0-fold > IM), respectively. They also analysed the level of neutralising antibodies in the sera 21 days post-vaccination and a few days prior to the challenge. Whilst neutralising antibody titres were relatively low in all animals, the OR (OR > IN > IM) produced the highest neutralising antibody titres. This study clearly highlighted the impact of immunisation with rVSV-ZEBOV on the induction of neutralising antibody responses [109]. Interestingly, NHPs previously infected with simian-human immunodeficiency virus were vaccinated with rVSV-ZEBOV. Following EBOV challenge, 4/6 animals survived. None of the six animals had a detectable antibody response by the day of challenge but three animals which survived developed a modest antibody response post-challenge, which suggests a role of antibodies in survival of these animals [110].

Recombinant VSV-ZEBOV (Erbevo®, Merck, Kenilworth, NJ, USA) is now the first fully licensed EBOV vaccine. This vaccine is delivered in humans in one dose and its safety and efficacy were also evaluated in clinical trials. Open-label, dose-escalation phase 1 trials were performed in healthy volunteers in Europe and Africa. They measured a persistent EBOV GP-specific antibody response in all vaccinated participants and higher neutralising responses when individuals received a higher dose of rVSV-ZEBOV vaccine [111]. Other phase 1 clinical trials reported very similar results in the US, Canada or Europe. In each study, a dose-dependent neutralising response was observed but occasionally neutralising responses failed to be detected in some individuals who received a lower dose of rVSV-ZEBOV vaccine [112–114]. One study provided clinical data on rVSV-ZEBOV efficacy, a phase 3 trial in Guinea, where a ring vaccination program was employed to include cases, contacts and contacts of contacts. The study reported 100% efficacy after 10 days and found that, 32 days following the detection of the first case in a vaccinated cluster, no new cases were reported, highlighting the ability of this vaccine to prevent transmission [115]. Despite the potential for bias relating to the standard of care between the experimental and control groups in the protocol of this study (as reported by some researchers), this latter highlights the high efficacy of the rVSV-ZEBOV vaccine in humans [116]. Halperin et al. showed that rVSV-ZEBOV produced strong antibody titres with 94% of participants becoming seropositive after 28 days and 91% remaining seropositive after 24 months. Geometric mean titres (GMT) rose from below the assay detection limit of 36.11 to 1262.0 after 28 days before gradually decreasing but retaining a high GMT of 920 after 24 months. The GMTs of neutralising antibodies, measured by PRNT, were high and continued to increase, peaking at 18 months with a plateau at 24 months [117]. Interestingly, Khurana et al. studied the human antibody repertoire in individuals vaccinated with rVSV-ZEBOV. They reported a high initial neutralising IgM immune response before IgG becomes the dominant subtype, which may explain the rapid protection provided by the vaccine. They demonstrated a higher diversity of antibody epitopes in vaccinees who received 20 million plaque-forming units (PFU) compared to those who received 3 or 100 million PFU. Another finding was that higher levels of neutralising GP-specific antibodies were induced after a single vaccination with 20 or 100 million PFU. A boost did not improve neutralising antibody response [118].

Adenoviral vector vaccines have been also developed in the context of EBOV. These vaccines are well known to induce robust T cell responses [119–121]. Wong et al. evaluated the antibody responses in the context of vaccine candidates based on AdHu5 expressing Zaire EBOV GP [105]. They compared the immune responses and survival rates of knockout mice (Rag-1$^{-/-}$, B cell$^{-/-}$, CD8^{+} T cell$^{-/-}$, IFN-$\gamma^{-/-}$, and CD4^{+} T cell$^{-/-}$), reporting that B cell and CD4^{+} T cell responses were the most critical in the development of a protective immune response against a mouse adapted strain of EBOV [105]. When repeated in guinea pigs, the average titres of anti-EBOV GP antibodies and neutralising antibody post-challenge were significantly higher in survivors than in non-survivors [105]. Interestingly, they also performed a similar study in NHPs. In this animal model, they found a higher anti-GP IgG response in survivors, but they did not detect any differences in neutralising antibody response between the survivors and non-survivors. However, a difference in T cell responses between the cohorts was detected in this latter model [105]. Chen et al. observed a durable EBOV-neutralising response in mice vaccinated with a prime-boost vaccine regimen based on a chimpanzee serotype 7 adenovirus expressing EBOV GP and a truncated version of EBOV GP1 protein [122].

In humans, the two-phase vaccine candidate Ad26.ZEBOV (Zabdeno®) and MVA-BN-Filo (Mvabea®), produced by Johnson&Johnson (New Brunswick, NJ, USA), is a vaccine candidate that has progressed well through clinical trials and has received marketing authorisation by the European Commission in July 2020. This is a viral vector vaccine, that uses two heterologous doses. The first dose is a non-replicating Adenovirus type-26 encoding EBOV GP and the second dose is a multivalent recombinant Modified Vaccinia Ankara vector-based vaccine encoding the GPs from EBOV, SUDV, MARV and the nucleoprotein of TAFV [123]. Anywaine et al. compared the immunogenicity of heterologous two-dose Ad26.ZEBOV and MVA-BN-Filo vaccination regimens in a phase 1 trial in Uganda and Tanzania. They demonstrated a robust immunogenicity when Ad26.ZEBOV was administered as the first dose followed by MVA-BN-Filo. This regimen promoted higher neutralising and total antibody titres with 93% of participants achieving seropositivity at the time of the second dose, reaching 100% 21 days after the second dose. Neutralising antibody titres were initially low but peaked at 21 days following the second dose 100% of participants in the 56-day interval group, before decreasing and stabilising 180 days post-initial vaccination. Titres plateaued until the final time-point at day 365 [124]. A similar phase 1 clinical trial was run in Kenya, where high levels of neutralising GP-specific antibodies were detected and sustained up to 360 days after the first dose [125].

Globally, preclinical and clinical studies demonstrated a key role of antibody responses in protection post-vaccination, particularly in the context of rVSV-ZEBOV. However, the correlation between neutralising responses and protection following vaccination is still debated.

Finally, there is uncertainty regarding the efficacy of postexposure mAb treatments following a recent vaccination with an EBOV vaccine. This scenario of exposure may happen early post-injection when vaccination is not yet fully effective. Cross et al. demonstrated in rhesus macaques that vaccination with rVSV-ZEBOV 1 day prior to EBOV challenge followed by anti-EBOV GP mAb MIL77 treatment 3 days later increased the rate of survival compared to animals vaccinated or treated with MIL77 only [126]. However, additional data is needed to draw a robust conclusion about a potential synergy between vaccination and antibody-based immunotherapy.

3.3.2. MARV Vaccine Candidates

To date no MARV vaccine has been approved yet, but several vaccine platforms have been attempted, including a recombinant VSV vector expressing MARV GP which showed efficacy in NHPs both post-exposure and prophylactically [127]. Jones et al. demonstrated that a single intramuscular injection led to 100% protection of vaccinated NHPs from a lethal MARV challenge [128]. Daddario-DiCaprio et al. confirmed these results with the same vaccine by proving the 100% protection following a challenge with heterologous MARV strains and RAVV [129]. Both studies reported high IgG titres in vaccinated NHPs but an absent or very low neutralising response. More recently, Mire et al. conducted a study involving six NHPs immunised with a rVSV-MARV-GP platform prior to challenge 13 months later. Immunisation elicited strong total antibody titres (between 1600 and 12,800) which remained elevated throughout the study. Regarding the neutralising antibody responses determined by PRNT, all NHPs had a neutralising response at day 28 post-vaccination but the titres were relatively low and by the day of challenge, two had no neutralising antibodies [130]. Together, these results suggest a less important role of neutralising antibodies in protection from MARV challenge.

Interestingly, a follow-up study by Daddario-DiCaprio et al. showed that immunisation with rVSV-MARV vaccine up to 20–30 min post-challenge protected all NHPs. All treated animals showed low to moderate amounts of IgM by day 6 post-challenge. Four of the five vaccinated animals developed a moderate IgG response by day 10 post-challenge. However, PRNT demonstrated low amounts of neutralising antibodies between days 6 and 37 in the plasma of all immunised animals [131]. Similar results were confirmed by Geisbert et al. In addition, a partial protection was reached when the vaccine was administered 24 h and 48 h post-challenge, when five out of six and two out of six macaques were protected, respectively. All animals which survived to challenge showed moderate to high levels of MARV-specific IgG in sera, whereas animals that died did not have any detectable IgG responses [132]. The sum of these results suggests the importance of humoral response following rVSV-MARV vaccination even though the neutralising activity may be less important with rVSV-MARV vaccine compared to rVSV-EBOV vaccine.

4. Methods to Measure Neutralising Antibody Responses

Researchers use different assays to evaluate neutralising antibody titres in serum or other biological fluids. To measure EBOV-specific neutralising antibody responses, the gold standard is the PRNT, which uses authentic live virus to measure the number of plaques formed upon infection of a cell line, such that if the sample is capable of neutralisation, fewer plaques would be observed [133]. Briefly the biological fluid (e.g., serum) containing the antibodies is mixed with the authentic live virus, usually for 1–2 h, and used to infect a permissive cell line. An agarose overlay is added for 5–7 days to avoid a 'too rapid' spreading of the virus. Finally, the cells can be fixed and plaques are counted, often manually. The PRNT takes several days and its development at a large-scale can be limited. It is the reason why microneutralisation assays (MNA) were also developed in the context of EBOV [97]. The concept of MNA is very similar to PRNT but 1 h post-incubation of cells with the mix EBOV/serum, the cells are washed and fixed for 1–2 days. Finally, infected cells can be detected using anti-EBOV and secondary antibodies. The number of spots can be counted using an imaging system [58]. Both methods, PRNT and MNA, are robust. However, EBOV is classified as a biosafety safety level (BSL) 4 organism as it is a very dangerous pathogen. BSL-4 work requires Class III safety cabinets, very specific pressure conditions in the lab, very strict decontamination procedures of the materials and trained staff. This increases the time and cost of such experiments, as well as limiting the number of sites with such capabilities [134].

Pseudotyped virus neutralisation assays can be a more flexible and manageable approach. Pseudotyped viruses (pseudoviruses) are recombinant viruses with core and envelope proteins derived from different viruses. They carry full or partial/modified sequence genomes to give rise to replication proficient or deficient virus. A reporter gene (e.g., luciferase, green fluorescent protein) is incorporated into the genome of the vector to detect a reduction of luminescence or fluorescence by neutralising antibodies, following infection of a permissive cell line. In the context of EBOV, recombinant VSV or lentiviral viruses are often used as a backbone to express EBOV GP. The advantage is that the recombinant pseudotyped viruses can usually be used at lower containment levels than EBOV. However, data generated from different pseudovirus systems can vary. Indeed, parameters such as the type of backbone, the reporter system and the expression of GP on the backbone may impact the sensitivity and the specificity of the assays, as well as their ability to accurately detect neutralising antibodies. Therefore, some studies have investigated conditions to improve the correlation with live EBOV neutralisation assays. We compared side by side two systems of EBOV GP pseudoviruses. Steeds et al. showed that the VSV luciferase pseudovirus system had a greater correlation (r = 0.85 + p < 0.0001) to the live EBOV assay than the HIV-1 system (r = 0.54 + p = 0.0004) [135,136]. Similarly, Wilkinson et al. ran a study across several laboratories and reported that labs using the VSV system reported higher correlation with a wildtype Ebola virus neutralisation assay (r = 0.84 and r = 0.96). Konduru et al. also confirmed a high correlation between the VSV platform and live EBOV neutralisation assay [137,138]. A limitation of the pseudovirus system is the absence of proteins aside from GP. While neutralisation antibodies mainly target EBOV GP, we cannot exclude the possibility that neutralising antibodies may be directed against other key proteins (e.g., VP40, other forms of GP) and cannot be detected using a pseudovirus system. For instance, VSV systems do not produce sGP, which could have an impact on neutralisation outcomes. Interestingly, Saphire et al. evaluated the impact of sGP on the results generated by three different neutralisation assays: VSV platform, authentic live EBOV, EBOVΔVP30-RenLUc virus. The VSV system did not express sGP whereas the other systems expressed wildtype sGP. Globally, they observed that the presence of sGP did not prevent neutralisation and in some cases, neutralisation was higher in the sGP-expressing assays [81]. When pseudoviruses are replication-deficient, this system can only detect the antibodies which prevent the entry of the virus into the cell. For example, the system cannot detect the impact of antibodies on the production of new viral progeny.

Pseudovirus platforms are surrogate systems to identify neutralising antibodies. However, the production of recombinant filoviruses expressing a reporter protein, which can help to monitor and quantify the infection, may be useful to improve the characterisation of antibodies able to neutralise EBOV or other filoviruses. Thus, an alternative approach to measuring filovirus-specific neutralising antibody responses is to use chimeric filoviruses. Chimeric filoviruses are man-made viruses composed of components from two filoviruses. Usually, one live filovirus is used as a backbone and its GP is replaced by a heterologous filovirus GP (e.g., EBOV GP). The potential difference in neutralisation-sensitivity between a chimeric filovirus and a pseudovirus has to be considered in order to avoid producing misleading results in the detection of neutralising antibodies [139]. Llinykh et al. compared the neutralisation-sensitivity of VSV-expressing filovirus GP with EBOV-expressing heterologous filovirus GP from BDBV, SUDV, MARV and LLOV. They reported that chimeric filoviruses were as sensitive as authentic filoviruses expressing the same GP. However, VSV chimeric viruses were more sensitive to antibody neutralisation than authentic filoviruses [139]. The analysis of neutralising antibody responses can also be performed by surrogate virus neutralisation tests. Whilst these tests have been broadly developed for other viruses such as SARS-CoV-2, to our knowledge they have yet to be developed for EBOV or other filoviruses.

To conclude, the choice of platform used to measure neutralising antibody responses is crucial in order to yield the most accurate results. However, the neutralising antibody titres generated with different platforms may be difficult to compare. It is the reason why the use of a WHO research standard in each assay can help to compare the titres between the studies and consequently, the neutralisation efficacy determined in each study.

5. Viral Neutralisation

5.1. Main Mechanisms of Neutralisation

Neutralising antibodies bind epitopes which block critical amino acids in the virus's lifecycle, typically epitopes involved in cell entry or the production of new viral progeny. The different mechanisms of neutralisation are shown in Figure 2 (pink boxes).

The vast majority of EBOV neutralising antibodies are targeted towards the GP as it contains multiple crucial epitopes exposed on the surface, particularly in the RBD [140]. Neutralising antibodies which bind to the EBOV GP block the interactions between the EBOV GP and NPC1 and prevent receptor binding via a direct binding to the RBD or in a way to block access to the RBD, thus preventing any subsequent binding [90]. The presence of neutralising antibodies can also prevent EBOV fusion with the endosomal membrane, usually necessary for the virus to escape into the cytoplasm. Indeed, neutralising antibodies can impede this by binding to the GP_1/GP_2 interface, clamping the GP and preventing the conformational changes in the GP required for this process (Figure 2) [17]. Finally, cathepsin cleavage usually removes the GC and MLD to expose the RBD mediating access for receptor binding. Neutralising antibodies can bind to GP in such a way to partially or totally prevent cleavage. For example, mAb CA45 disrupts cleavage to produce partially cleaved GPs, while mAb100 almost completely inhibits cleavage [54,140,141]. Though not clearly described in the context of EBOV, some antibodies can neutralise by altering the structure or conformation of the viral GP to the point that it is unable to perform a critical function, e.g., receptor binding (see Figure 2) [142]. VP40 is critical for viral budding and the production of new progeny and therefore its potential as a therapeutic target has been investigated. For example, anti-VP40 mAbs have been shown to neutralise EBOV by preventing the formation of new progeny (Figure 2). This is bolstered when administered in cocktails with mAbs targeting different sites on the VP40 protein or with GP-specific antibodies [143,144]. Antibodies, particularly polymeric IgM and IgA, can also agglutinate virus particles, forming aggregates that reduce the contact between the virus and the cell and affect subsequent entry, in addition to signalling, for antibody-dependent phagocytosis (Figure 2) [145–147].

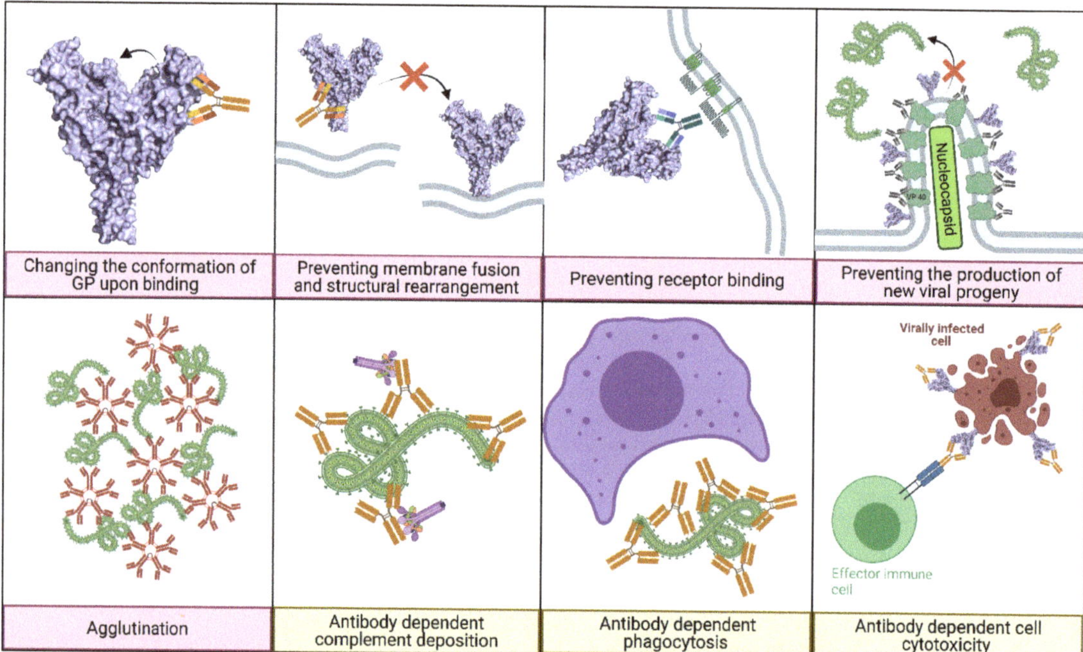

Figure 2. The mechanisms of action of neutralising antibodies (pink boxes) and non-neutralising antibodies (yellow boxes). Diagram was created with BioRender.com (©BioRender 2021, accessed in June 2021) and the protein structure was generated in The PyMOL Molecular Graphics System, Version 2.0 Schrödinger, Delano Scientific LLC, Berkeley, CA, USA using the PDB accession: 5JQ3.

Finally, it is important to highlight that based on some studies performed by the Viral Haemorrhagic Fever Immunotherapeutic Consortium, 5% of non- or weakly neutralising EBOV-specific antibodies tested in the panel were found to be protective when assessed in vivo proving a crucial role for non-neutralising antibodies in protection [81]. Perhaps more protective antibodies have not yet been discovered, as the assays of choice for the analysis of antibody responses are largely based on neutralisation only. Indeed, non-neutralising antibodies can stimulate an array of processes that aid in viral clearance including initiating complement deposition, contributing to viral inactivation, phagocytosis and complement-dependent cytotoxicity (Figure 2) [148]. Monoclonal antibodies are also capable of interacting with Fc receptors activating antibody-dependent cell-mediated cytotoxicity and antibody-dependent cellular phagocytosis [68,146,148,149]. Furthermore, it has been reported that non-neutralising antibodies have been shown to have synergistic neutralising ability, whereby they bind to the viral GP, altering the structure of an epitope making it more accessible for neutralising antibodies to bind. This has been documented for both EBOV and SUDV [150]. The three main mechanisms of non-neutralising antibodies are mentioned in Figure 2 (yellow boxes). Some monoclonal antibodies isolated from EBOV survivors have been shown, at sub-neutralising concentrations, to cause antibody-dependent enhancement via antibody-dependent phagocytosis, especially in macrophages which are thought to be the preferred target cell of EBOV. This could play a potential role in antibody-dependent enhancement of EBOV infection leading to a more severe disease and worse clinical outcomes [151].

5.2. Regions Targeted by Neutralising Antibodies

Antibodies binding to the GP head are frequently neutralising because they block the function of the RBD, hence inhibiting cell entry and GP binding to NPC1 which is required for escaping the endosome.

The Viral Haemorrhagic Fever Immunotherapeutic Consortium investigated 168 EBOV specific mAbs for their ability to neutralise live virus and pseudotyped VSV and measured their ability to protect BALB/c mice following a challenge with mouse-adapted EBOV. The study found that antibodies specific for the GP_1 head, base, the HR2 and IFL provided statistically significant protection. All of the same domains, except the GP_1 base were targeted by more potently neutralising antibodies. Consequently, the neutralising ability of the antibodies showed a statistically significant correlation with protection, with a Spearman's rank value of 0.65 (the closer rank value is to 1, the stronger the association between the ranks), based on the percentage of infected cells using an authentic EBOV [81]. HR1, HR2 and IFL are also important neutralisation targets given that HR1 and HR2 both facilitate IFL insertion into the host membrane [12]. Some studies clearly described that the neutralising activity of CA45 or ADI-15878 and ADI-15742 were linked to IFL binding [58,93]. Finally, the base of GP_1 is also a target for neutralisation, when an antibody binds, it prevents the release of the base which clamps HR1 and HR2 to instigate the conformational changes required for viral escape from the endosome, thereby preventing these conformational changes from taking place [12].

The GC and MLD are designed to shield GP from humoral responses and are cleaved by cathepsin in the endosome. They play no part in the receptor binding or fusion of the EBOV GP with the cell membrane, hence are not always targets for neutralisation [81]. This is demonstrated by the potently neutralising mAb114 and the non-neutralising 13C6 which have overlapping epitopes binding to the GP_1 core and at least partially to the GC. MAb114's binding to the GC isn't critical and remains bound and neutralising when the GC is removed following cathepsin cleavage. Meanwhile, residues on GC are critical for 13C6 binding and when removed, 13C6 no longer remains bound to the EBOV GP, rendering it non-neutralising and allowing EBOV to freely interact with NPC1 [75].

6. Strategies to Prevent Immune Escape from Neutralising Antibodies

The efficacy of antibody-based therapeutics and vaccine-induced antibody responses may be reduced over time due to viral immune evasion. Immune evasion may impair antibody binding and consequently neutralisation reducing protection provided by the mAbs-based therapeutics, vaccines or a previous natural response.

A major drawback of mAbs is that they are specific to one epitope. Consequently, they are more vulnerable to immune escape from rapidly mutating viruses, and this has been observed with many other viruses such as influenza, HIV or SARS-CoV-2 [77,152–155]. In an experimental setting, Steeds et al. showed that GP variants G74R, P330S and H407Y were capable of readily escaping the WHO research neutralisation standard KZ52. Worryingly, KZ52 has overlapping epitopes with several other promising mAb therapeutics, e.g., CA45, 4G7 and more (Table 1) [135]. A solution to this concern is the use of mAb cocktails, that target several distinct epitopes, hence reducing the chance of immune evasion. It is the reason why numerous mAbs discussed previously are part of a cocktail. Furthermore, this strategy also allows the addition of non-neutralising antibodies which may be more effective at initiating effector functions to increase the efficacy of the drug e.g., 13C6 or Odesivimab [77,80]. Another tactic that is employed involves targeting highly conserved epitopes that are often critical to viral fitness.

Currently, there are only two licensed EBOV-specific mAbs, with no licensed therapeutics on the market for the treatment of any other filovirus. Thus, pan-ebolavirus or pan-filovirus mAbs would significantly improve filovirus treatment options. An example of which includes CA45 and FVM04 that when used in combination as a cocktail provide 100% protection against both EBOV and SUDV in NHP and mouse models [87,93]. Furthermore, when MR191, is added to the cocktail, it also provides 100% protection against MARV [87]. Interestingly, some pan-neutralising mAbs were isolated from a 2013–2016 West Africa EBOV epidemic survivor and showed protection in mouse and ferret models [58]. Gilchuk et al. also isolated mAbs from the 2013–2016 West African epidemic and 2018 DRC outbreak survivors targeting epitopes at the base region of the GP which displayed pan-neutralising and protective abilities in mice, guinea pigs and ferrets [83]. Some survivors from the 2014 Boende EVD outbreak also mounted pan-filovirus serum neutralising responses [156].

There has also recently been a push for cross-reactive pan-filovirus vaccine candidates. Experimental DNA- and VSV-based vaccines encoding multiple GPs of various filoviruses have been tested in NHP models. For example, Keck et al. immunised two cynomolgus macaques three times with a trivalent GP cocktail consisting of EBOV/SUDV/MARV. Sera from the macaques had strong IgG measured by ELISA. While neutralising titres against the GP of all three filoviruses, were detected, although significantly lower neutralisation was observed with MARV. As MARV is more distantly related, it is not surprising that less cross-reactivity is shared [157].

The development of pan-neutralising therapeutics and vaccines is likely the future of filovirus research. These therapies may prove to be a highly beneficial resource in the poor regions where filoviruses are endemic.

7. Conclusions

This review has summarised that antibodies, and in particular neutralising antibodies, are an important line of defence against filovirus infection but current research falls short of defining them as a correlate of protection. Preclinical and clinical studies were not able to determine the level of neutralising antibodies necessary to protect an individual.

Monoclonal antibodies are the first licensed post-exposure therapeutics for EVD, which, coupled with expanded use of vaccines, have the potential to save a large number of lives. However, vaccine access in the field is still limited and mAb therapies are not perfect. In addition, the emergence of EBOV variants which could reduce the efficacy of mAbs and vaccine-induced humoral responses has to be considered. Consequently, newer and more effective antibody-based treatments containing more broadly neutralising antibodies, as well as the development of vaccines based on epitopes eliciting cross-reactive neutralising antibodies, would be very beneficial.

Author Contributions: A.H. and S.L. wrote the manuscript and designed the figures. C.B., J.M., T.T. and M.W.C. revised the manuscript. All authors contributed to the article and approved the submitted version. All authors have read and agreed to the published version of the manuscript.

Funding: The APC was funded by the US Food and Drug Administration (grant number: 75F40120C 00085).

Institutional Review Board Statement: Not applicable.

Informed Consent Statement: Not applicable.

Conflicts of Interest: The authors declare no conflict of interest.

References

1. Towner, J.S.; Khristova, M.L.; Sealy, T.K.; Vincent, M.J.; Erickson, B.R.; Bawiec, D.A.; Hartman, A.; Comer, J.A.; Zaki, S.R.; Ströher, U.; et al. Marburgvirus Genomics and Association with a Large Hemorrhagic Fever Outbreak in Angola. *J. Virol.* **2006**, *80*, 6497–6516. [CrossRef] [PubMed]
2. Timothy, J.; Hall, Y.; Akoi-Boré, J.; Diallo, B.; Tipton, T.R.W.; Bower, H.; Strecker, T.; Glynn, J.R.; Carroll, M. Early transmission and case fatality of Ebola virus at the index site of the 2013–2016 west African Ebola outbreak: A cross-sectional seroprevalence survey. *Lancet Infect. Dis.* **2019**, *19*, 429–438. [CrossRef]
3. Goldstein, T.; Anthony, S.J.; Gbakima, A.; Bird, B.H.; Bangura, J.; Tremeau-Bravard, A.; Belaganahalli, M.N.; Wells, H.L.; Dhanota, J.K.; Liang, E.; et al. The discovery of Bombali virus adds further support for bats as hosts of ebolaviruses. *Nat. Microbiol.* **2018**, *3*, 1084–1089. [CrossRef]
4. Zampieri, C.A.; Sullivan, N.J.; Nabel, G.J. Immunopathology of highly virulent pathogens: Insights from Ebola virus. *Nat. Immunol.* **2007**, *8*, 1159–1164. [CrossRef] [PubMed]
5. Negredo, A.; Palacios, G.; Vázquez-Morón, S.; González, F.; Dopazo, H.; Molero, F.; Juste, J.; Quetglas, J.; Savji, N.; Martínez, M.D.L.C.; et al. Discovery of an Ebolavirus-Like Filovirus in Europe. *PLoS Pathog.* **2011**, *7*, e1002304. [CrossRef]
6. Ramírez de Arellano, E.; Sanchez-Lockhart, M.; Perteguer, M.J.; Bartlett, M.; Ortiz, M.; Campioli, P.; Hernández, A.; Gonzalez, J.; Garcia, K.; Ramos, M.; et al. First Evidence of Antibodies against Lloviu Virus in Schreiber's Bent-Winged Insectivorous Bats Demonstrate a Wide Circulation of the Virus in Spain. *Viruses* **2019**, *11*, 360. [CrossRef]
7. Kemenesi, G.; Kurucz, K.; Dallos, B.; Zana, B.; Földes, F.; Boldogh, S.; Görföl, T.; Carroll, M.W.; Jakab, F. Re-emergence of Lloviu virus in Miniopterus schreibersii bats, Hungary, 2016. *Emerg. Microbes Infect.* **2018**, *7*, 1–4. [CrossRef]
8. Yang, X.-L.; Tan, C.W.; Anderson, D.E.; Jiang, R.-D.; Li, B.; Zhang, W.; Zhu, Y.; Lim, X.F.; Zhou, P.; Liu, X.-L.; et al. Characterization of a filovirus (Měnglà virus) from Rousettus bats in China. *Nat. Microbiol.* **2019**, *4*, 390–395. [CrossRef]
9. Burk, R.; Bollinger, L.; Johnson, J.; Wada, J.; Radoshitzky, S.; Palacios, G.; Bavari, S.; Jahrling, P.B.; Kuhn, J.H. Neglected filoviruses. *FEMS Microbiol. Rev.* **2016**, *40*, 494–519. [CrossRef]
10. Maruyama, J.; Miyamoto, H.; Kajihara, M.; Ogawa, H.; Maeda, K.; Sakoda, Y.; Yoshida, R.; Takada, A. Characterization of the Envelope Glycoprotein of a Novel Filovirus, Lloviu Virus. *J. Virol.* **2014**, *88*, 99–109. [CrossRef]
11. Hume, A.; Mühlberger, E. Distinct Genome Replication and Transcription Strategies within the Growing Filovirus Family. *J. Mol. Biol.* **2019**, *431*, 4290–4320. [CrossRef]
12. Lee, J.E.; Saphire, E.O. Ebolavirus glycoprotein structure and mechanism of entry. *Future Virol.* **2009**, *4*, 621–635. [CrossRef] [PubMed]
13. Zhao, Y.; Ren, J.; Harlos, K.; Jones, D.M.; Zeltina, A.; Bowden, T.A.; Padilla-Parra, S.; Fry, E.E.; Stuart, D.I. Toremifene interacts with and destabilizes the Ebola virus glycoprotein. *Nature* **2016**, *535*, 169–172. [CrossRef] [PubMed]
14. Mühlberger, E. Filovirus replication and transcription. *Future Virol.* **2007**, *2*, 205–215. [CrossRef] [PubMed]
15. Cantoni, D.; Rossman, J.S. Ebolaviruses: New roles for old proteins. *PLoS Negl. Trop. Dis.* **2018**, *12*, e0006349. [CrossRef]
16. Volchkov, V.E.; Feldmann, H.; Volchkova, V.A.; Klenk, H.-D. Processing of the Ebola virus glycoprotein by the proprotein convertase furin. *Proc. Natl. Acad. Sci. USA* **1998**, *95*, 5762–5767. [CrossRef] [PubMed]
17. Lee, J.E.; Fusco, M.L.; Hessell, A.J.; Oswald, W.B.; Burton, D.R.; Saphire, E.O. Structure of the Ebola virus glycoprotein bound to an antibody from a human survivor. *Nature* **2008**, *454*, 177–182. [CrossRef] [PubMed]
18. Zhu, W.; Banadyga, L.; Emeterio, K.; Wong, G.; Qiu, X. The Roles of Ebola Virus Soluble Glycoprotein in Replication, Pathogenesis, and Countermeasure Development. *Viruses* **2019**, *11*, 999. [CrossRef] [PubMed]
19. Mehedi, M.; Falzarano, D.; Seebach, J.; Hu, X.; Carpenter, M.S.; Schnittler, H.-J.; Feldmann, H. A New Ebola Virus Nonstructural Glycoprotein Expressed through RNA Editing. *J. Virol.* **2011**, *85*, 5406–5414. [CrossRef]
20. Escudero-Pérez, B.; Volchkova, V.A.; Dolnik, O.; Lawrence, P.; Volchkov, V.E. Shed GP of Ebola Virus Triggers Immune Activation and Increased Vascular Permeability. *PLoS Pathog.* **2014**, *10*, e1004509. [CrossRef]
21. Yu, D.-S.; Weng, T.-H.; Wu, X.; Wang, F.X.; Lu, X.-Y.; Wu, H.-B.; Wu, N.-P.; Li, L.-J.; Yao, H.-P. The lifecycle of the Ebola virus in host cells. *Oncotarget* **2017**, *8*, 55750–55759. [CrossRef] [PubMed]
22. Nanbo, A.; Imai, M.; Watanabe, S.; Noda, T.; Takahashi, K.; Neumann, G.; Halfmann, P.; Kawaoka, Y. Ebolavirus Is Internalized into Host Cells via Macropinocytosis in a Viral Glycoprotein-Dependent Manner. *PLoS Pathog.* **2010**, *6*, e1001121. [CrossRef] [PubMed]
23. Brecher, M.; Schornberg, K.L.; Delos, S.E.; Fusco, M.L.; Saphire, E.O.; White, J.M. Cathepsin Cleavage Potentiates the Ebola Virus Glycoprotein to Undergo a Subsequent Fusion-Relevant Conformational Change. *J. Virol.* **2012**, *86*, 364–372. [CrossRef] [PubMed]
24. Miller, E.H.; Obernosterer, G.; Raaben, M.; Herbert, A.S.; Deffieu, M.S.; Krishnan, A.; Ndungo, E.; Sandesara, R.G.; Carette, J.; Kuehne, A.I.; et al. Ebola virus entry requires the host-programmed recognition of an intracellular receptor. *EMBO J.* **2012**, *31*, 1947–1960. [CrossRef] [PubMed]
25. Gregory, S.M.; Larsson, P.; Nelson, E.A.; Kasson, P.M.; White, J.M.; Tamm, L.K. Ebolavirus Entry Requires a Compact Hydrophobic Fist at the Tip of the Fusion Loop. *J. Virol.* **2014**, *88*, 6636–6649. [CrossRef]
26. Moller-Tank, S.; Maury, W. Ebola Virus Entry: A Curious and Complex Series of Events. *PLoS Pathog.* **2015**, *11*, e1004731. [CrossRef]
27. Olejnik, J.; Ryabchikova, E.; Corley, R.B.; Mühlberger, E. Intracellular Events and Cell Fate in Filovirus Infection. *Viruses* **2011**, *3*, 1501–1531. [CrossRef]

28. Okumura, A.; Pitha, P.M.; Yoshimura, A.; Harty, R.N. Interaction between Ebola Virus Glycoprotein and Host Toll-Like Receptor 4 Leads to Induction of Proinflammatory Cytokines and SOCS1. *J. Virol.* **2010**, *84*, 27–33. [CrossRef]
29. Olejnik, J.; Forero, A.; Deflubé, L.R.; Hume, A.; Manhart, W.A.; Nishida, A.; Marzi, A.; Katze, M.G.; Ebihara, H.; Rasmussen, A.L.; et al. Ebolaviruses Associated with Differential Pathogenicity Induce Distinct Host Responses in Human Macrophages. *J. Virol.* **2017**, *91*, e00179-17. [CrossRef]
30. Gupta, M.; Mahanty, S.; Ahmed, R.; Rollin, P. Monocyte-Derived Human Macrophages and Peripheral Blood Mononuclear Cells Infected with Ebola Virus Secrete MIP-1α and TNF-α and Inhibit Poly-IC-Induced IFN-α in Vitro. *Virology* **2001**, *284*, 20–25. [CrossRef]
31. Ströher, U.; West, E.; Bugany, H.; Klenk, H.-D.; Schnittler, H.-J.; Feldmann, H. Infection and Activation of Monocytes by Marburg and Ebola Viruses. *J. Virol.* **2001**, *75*, 11025–11033. [CrossRef]
32. Hensley, L.E.; Young, H.A.; Jahrling, P.B.; Geisberta, T.W. Proinflammatory response during Ebola virus infection of primate models: Possible involvement of the tumor necrosis factor receptor superfamily. *Immunol. Lett.* **2002**, *80*, 169–179. [CrossRef]
33. Wahl-Jensen, V.; Kurz, S.; Feldmann, F.; Buehler, L.K.; Kindrachuk, J.; De Filippis, V.; Correia, J.D.S.; Früh, K.; Kuhn, J.H.; Burton, D.R.; et al. Ebola Virion Attachment and Entry into Human Macrophages Profoundly Effects Early Cellular Gene Expression. *PLoS Negl. Trop. Dis.* **2011**, *5*, e1359. [CrossRef] [PubMed]
34. McElroy, A.K.; Muhlberger, E.; Muñoz-Fontela, C. Immune barriers of Ebola virus infection. *Curr. Opin. Virol.* **2018**, *28*, 152–160. [CrossRef] [PubMed]
35. Longhi, P.; Trumpfheller, C.; Idoyaga, J.; Caskey, M.; Matos, I.; Kluger, C.; Salazar, A.M.; Colonna, M.; Steinman, R.M. Dendritic cells require a systemic type I interferon response to mature and induce CD4$^+$ Th1 immunity with poly IC as adjuvant. *J. Exp. Med.* **2009**, *206*, 1589–1602. [CrossRef] [PubMed]
36. Rhein, B.A.; Powers, L.S.; Rogers, K.; Anantpadma, M.; Singh, B.; Sakurai, Y.; Bair, T.; Miller-Hunt, C.; Sinn, P.; Davey, R.; et al. Interferon-γ Inhibits Ebola Virus Infection. *PLoS Pathog.* **2015**, *11*, e1005263. [CrossRef]
37. Dyall, J.; Hart, B.J.; Postnikova, E.; Cong, Y.; Zhou, H.; Gerhardt, D.M.; Freeburger, D.; Michelotti, J.; Honko, A.N.; Dewald, L.E.; et al. Interferon-β and Interferon-γ Are Weak Inhibitors of Ebola Virus in Cell-Based Assays. *J. Infect. Dis.* **2017**, *215*, 1416–1420. [CrossRef] [PubMed]
38. Jahrling, P.B.; Geisbert, T.W.; Geisbert, J.B.; Swearengen, J.R.; Bray, M.; Jaax, N.K.; Huggins, J.W.; LeDuc, J.W.; Peters, C.J. Evaluation of Immune Globulin and Recombinant Interferon-α2b for Treatment of Experimental Ebola Virus Infections. *J. Infect. Dis.* **1999**, *179*, S224–S234. [CrossRef]
39. Smith, L.M.; Hensley, L.; Geisbert, T.W.; Johnson, J.; Stossel, A.; Honko, A.; Yen, J.Y.; Geisbert, J.; Paragas, J.; Fritz, E.; et al. Interferon-β Therapy Prolongs Survival in Rhesus Macaque Models of Ebola and Marburg Hemorrhagic Fever. *J. Infect. Dis.* **2013**, *208*, 310–318. [CrossRef] [PubMed]
40. Qiu, X.; Wong, G.; Fernando, L.; Audet, J.; Bello, A.; Strong, J.; Alimonti, J.B.; Kobinger, G.P. mAbs and Ad-Vectored IFN- Therapy Rescue Ebola-Infected Nonhuman Primates When Administered After the Detection of Viremia and Symptoms. *Sci. Transl. Med.* **2013**, *5*, 207ra143. [CrossRef]
41. Villinger, F.; Rollin, P.; Brar, S.S.; Chikkala, N.F.; Winter, J.; Sundstrom, J.B.; Zaki, S.R.; Swanepoel, R.; Ansari, A.A.; Peters, C.J. Markedly Elevated Levels of Interferon (IFN)-σ, IFN-α, Interleukin (IL)-2, IL-10, and Tumor Necrosis Factor-α Associated with Fatal Ebola Virus Infection. *J. Infect. Dis.* **1999**, *179*, S188–S191. [CrossRef] [PubMed]
42. Olejnik, J.; Hume, A.J.; Leung, D.W.; Amarasinghe, G.K.; Basler, C.F.; Mühlberger, E. Filovirus Strategies to Escape Antiviral Responses. *Curr. Top. Microbiol. Immunol.* **2017**, *411*, 293–322. [CrossRef]
43. McElroy, A.; Akondy, R.; Davis, C.W.; Ellebedy, A.H.; Mehta, A.; Kraft, C.S.; Lyon, G.M.; Ribner, B.S.; Varkey, J.; Sidney, J.; et al. Human Ebola virus infection results in substantial immune activation. *Proc. Natl. Acad. Sci. USA* **2015**, *112*, 4719–4724. [CrossRef]
44. Ruibal, P.; Oestereich, L.; Lüdtke, A.; Becker-Ziaja, B.; Wozniak, D.; Kerber, R.; Korva, M.; Cabeza-Cabrerizo, M.; Bore, J.A.; Koundouno, F.R.; et al. Unique human immune signature of Ebola virus disease in Guinea. *Nature* **2016**, *533*, 100–104. [CrossRef] [PubMed]
45. Baize, S.; Leroy, E.; Georges-Courbot, M.-C.; Capron, M.; Lansoud-Soukate, J.; Debré, P.; Fisher-Hoch, S.P.; McCormick, J.B.; Georges, A.J. Defective humoral responses and extensive intravascular apoptosis are associated with fatal outcome in Ebola virus-infected patients. *Nat. Med.* **1999**, *5*, 423–426. [CrossRef]
46. Muñoz-Fontela, C.; McElroy, A.K. Ebola Virus Disease in Humans: Pathophysiology and Immunity. *Curr. Top. Microbiol. Immunol.* **2017**, *411*, 141–169. [CrossRef]
47. Gupta, M.; Mahanty, S.; Greer, P.; Towner, J.S.; Shieh, W.-J.; Zaki, S.R.; Ahmed, R.; Rollin, P.E. Persistent Infection with Ebola Virus under Conditions of Partial Immunity. *J. Virol.* **2004**, *78*, 958–967. [CrossRef] [PubMed]
48. Thom, R.; Tipton, T.; Strecker, T.; Hall, Y.; Bore, J.A.; Maes, P.; Koundouno, F.R.; Fehling, S.K.; Krähling, V.; Steeds, K.; et al. Longitudinal antibody and T cell responses in Ebola virus disease survivors and contacts: An observational cohort study. *Lancet Infect. Dis.* **2020**, *21*, 507–516. [CrossRef]
49. Tipton, T.R.W.; Hall, Y.; Bore, J.A.; White, A.; Sibley, L.S.; Sarfas, C.; Yuki, Y.; Martin, M.; Longet, S.; Mellors, J.; et al. Characterisation of the T-cell response to Ebola virus glycoprotein amongst survivors of the 2013–16 West Africa epidemic. *Nat. Commun.* **2021**, *12*, 1153. [CrossRef] [PubMed]

50. Sakabe, S.; Sullivan, B.; Hartnett, J.N.; Robles-Sikisaka, R.; Gangavarapu, K.; Cubitt, B.; Ware, B.C.; Kotliar, D.; Branco, L.M.; Goba, A.; et al. Analysis of CD8$^+$ T cell response during the 2013–2016 Ebola epidemic in West Africa. *Proc. Natl. Acad. Sci. USA* **2018**, *115*, E7578–E7586. [CrossRef] [PubMed]

51. Longet, S.; Mellors, J.; Carroll, M.W.; Tipton, T. Ebolavirus: Comparison of Survivor Immunology and Animal Models in the Search for a Correlate of Protection. *Front. Immunol.* **2021**, *11*, 599568. [CrossRef]

52. Ksiazek, T.G.; Rollin, P.; Williams, A.J.; Bressler, D.S.; Martin, M.L.; Swanepoel, R.; Burt, F.J.; Leman, P.A.; Khan, A.S.; Rowe, A.K.; et al. Clinical Virology of Ebola Hemorrhagic Fever (EHF): Virus, Virus Antigen, and IgG and IgM Antibody Findings among EHF Patients in Kikwit, Democratic Republic of the Congo, 1995. *J. Infect. Dis.* **1999**, *179*, S177–S187. [CrossRef]

53. Song, P.; Zheng, N.; Liu, Y.; Tian, C.; Wu, X.; Ma, X.; Chen, D.; Zou, X.; Wang, G.; Wang, H.; et al. Deficient humoral responses and disrupted B-cell immunity are associated with fatal SFTSV infection. *Nat. Commun.* **2018**, *9*, 3328. [CrossRef]

54. Corti, D.; Misasi, J.; Mulangu, S.; Stanley, D.A.; Kanekiyo, M.; Wollen, S.; Ploquin, A.; Doria-Rose, N.A.; Staupe, R.P.; Bailey, M.; et al. Protective monotherapy against lethal Ebola virus infection by a potently neutralizing antibody. *Science* **2016**, *351*, 1339–1342. [CrossRef] [PubMed]

55. Maruyama, T.; Rodriguez, L.L.; Jahrling, P.B.; Sanchez, A.; Khan, A.S.; Nichol, S.T.; Peters, C.J.; Parren, P.W.H.I.; Burton, D.R. Ebola Virus Can Be Effectively Neutralized by Antibody Produced in Natural Human Infection. *J. Virol.* **1999**, *73*, 6024–6030. [CrossRef]

56. Parren, P.; Geisbert, T.W.; Maruyama, T.; Jahrling, P.B.; Burton, D.R. Pre- and Postexposure Prophylaxis of Ebola Virus Infection in an Animal Model by Passive Transfer of a Neutralizing Human Antibody. *J. Virol.* **2002**, *76*, 6408–6412. [CrossRef] [PubMed]

57. Flyak, A.; Shen, X.; Murin, C.D.; Turner, H.L.; David, J.; Fusco, M.L.; Lampley, R.; Kose, N.; Ilinykh, P.A.; Kuzmina, N.; et al. Cross-Reactive and Potent Neutralizing Antibody Responses in Human Survivors of Natural Ebolavirus Infection. *Cell* **2016**, *164*, 392–405. [CrossRef] [PubMed]

58. Wec, A.Z.; Herbert, A.S.; Murin, C.D.; Nyakatura, E.K.; Abelson, D.M.; Fels, J.M.; He, S.; James, R.M.; de La Vega, M.-A.; Zhu, W.; et al. Antibodies from a Human Survivor Define Sites of Vulnerability for Broad Protection against Ebolaviruses. *Cell* **2017**, *169*, 878–890.e15. [CrossRef] [PubMed]

59. Bornholdt, Z.; Turner, H.L.; Murin, C.D.; Li, W.; Sok, D.; Souders, C.A.; Piper, A.E.; Goff, A.; Shamblin, J.D.; Wollen-Roberts, S.; et al. Isolation of potent neutralizing antibodies from a survivor of the 2014 Ebola virus outbreak. *Science* **2016**, *351*, 1078–1083. [CrossRef] [PubMed]

60. Saphire, E.O.; Schendel, S.; Gunn, B.M.; Milligan, J.C.; Alter, G. Antibody-mediated protection against Ebola virus. *Nat. Immunol.* **2018**, *19*, 1169–1178. [CrossRef] [PubMed]

61. Klingler, J.; Weiss, S.; Itri, V.; Liu, X.; Oguntuyo, K.Y.; Stevens, C.; Ikegame, S.; Hung, C.-T.; Enyindah-Asonye, G.; Amanat, F.; et al. Role of IgM and IgA Antibodies in the Neutralization of SARS-CoV-2. *medRxiv* **2020**. [CrossRef]

62. Luczkowiak, J.; Arribas, J.R.; Gómez, S.; Jiménez-Yuste, V.; de la Calle-Prieto, F.; Viejo, A.; Delgado, R. Specific neutralizing response in plasma from convalescent patients of Ebola Virus Disease against the West Africa Makona variant of Ebola virus. *Virus Res.* **2016**, *213*, 224–229. [CrossRef]

63. Williamson, L.E.; Flyak, A.I.; Kose, N.; Bombardi, R.; Branchizio, A.; Reddy, S.; Davidson, E.; Doranz, B.J.; Fusco, M.L.; Saphire, E.O.; et al. Early Human B Cell Response to Ebola Virus in Four U.S. Survivors of Infection. *J. Virol.* **2019**, *93*. [CrossRef] [PubMed]

64. Rimoin, A.W.; Lu, K.; Bramble, M.S.; Steffen, I.; Doshi, R.H.; Hoff, N.A.; Mukadi, P.; Nicholson, B.P.; Alfonso, V.H.; Olinger, G.; et al. Ebola Virus Neutralizing Antibodies Detectable in Survivors of the Yambuku, Zaire Outbreak 40 Years after Infection. *J. Infect. Dis.* **2018**, *217*, 223–231. [CrossRef] [PubMed]

65. Halfmann, P.J.; Eisfeld, A.J.; Watanabe, T.; Maemura, T.; Yamashita, M.; Fukuyama, S.; Armbrust, T.; Rozich, I.; N'Jai, A.; Neumann, G.; et al. Serological analysis of Ebola virus survivors and close contacts in Sierra Leone: A cross-sectional study. *PLoS Negl. Trop. Dis.* **2019**, *13*, e0007654. [CrossRef]

66. Davis, C.W.; Jackson, K.; McElroy, A.K.; Halfmann, P.; Huang, J.; Chennareddy, C.; Piper, A.E.; Leung, Y.; Albariño, C.G.; Crozier, I.; et al. Longitudinal Analysis of the Human B Cell Response to Ebola Virus Infection. *Cell* **2019**, *177*, 1566–1582.e17. [CrossRef] [PubMed]

67. Khurana, S.; Ravichandran, S.; Hahn, M.; Coyle, E.M.; Stonier, S.W.; Zak, S.E.; Kindrachuk, J.; Davey, R.T.; Dye, J.M.; Chertow, D.S. Longitudinal Human Antibody Repertoire against Complete Viral Proteome from Ebola Virus Survivor Reveals Protective Sites for Vaccine Design. *Cell Host Microbe* **2020**, *27*, 262–276.e4. [CrossRef]

68. Gunn, B.M.; Roy, V.; Karim, M.M.; Hartnett, J.N.; Suscovich, T.J.; Goba, A.; Momoh, M.; Sandi, J.D.; Kanneh, L.; Andersen, K.G.; et al. Survivors of Ebola Virus Disease Develop Polyfunctional Antibody Responses. *J. Infect. Dis.* **2020**, *221*, 156–161. [CrossRef]

69. Ullah, I.; Prévost, J.; Ladinsky, M.S.; Stone, H.; Lu, M.; Anand, S.P.; Beaudoin-Bussières, G.; Symmes, K.; Benlarbi, M.; Ding, S.; et al. Live Imaging of SARS-CoV-2 Infection in Mice Reveals that Neutralizing Antibodies Require Fc Function for Optimal Efficacy. *Immunity* **2021**. [CrossRef] [PubMed]

70. Sobarzo, A.; Groseth, A.; Dolnik, O.; Becker, S.; Lutwama, J.J.; Perelman, E.; Yavelsky, V.; Muhammad, M.; Kuehne, A.I.; Marks, R.S.; et al. Profile and Persistence of the Virus-Specific Neutralizing Humoral Immune Response in Human Survivors of Sudan Ebolavirus (Gulu). *J. Infect. Dis.* **2013**, *208*, 299–309. [CrossRef] [PubMed]

71. Sobarzo, A.; Eskira, Y.; Herbert, A.S.; Kuehne, A.I.; Stonier, S.W.; Ochayon, D.E.; Fedida-Metula, S.; Balinandi, S.; Kislev, Y.; Tali, N.; et al. Immune Memory to Sudan Virus: Comparison between Two Separate Disease Outbreaks. *Viruses* **2015**, *7*, 37–51. [CrossRef]

72. Stonier, S.W.; Herbert, A.S.; Kuehne, A.I.; Sobarzo, A.; Habibulin, P.; Dahan, C.V.A.; James, R.M.; Egesa, M.; Cose, S.; Lutwama, J.J.; et al. Marburg virus survivor immune responses are Th1 skewed with limited neutralizing antibody responses. *J. Exp. Med.* **2017**, *214*, 2563–2572. [CrossRef]

73. Lipman, N.S.; Jackson, L.R.; Trudel, L.J.; Weis-Garcia, F. Monoclonal Versus Polyclonal Antibodies: Distinguishing Characteristics, Applications, and Information Resources. *ILAR J.* **2005**, *46*, 258–268. [CrossRef]

74. Gaudinski, M.; Coates, E.E.; Novik, L.; Widge, A.; Houser, K.V.; Burch, E.; Holman, L.A.; Gordon, I.J.; Chen, G.L.; Carter, C.; et al. Safety, tolerability, pharmacokinetics, and immunogenicity of the therapeutic monoclonal antibody mAb114 targeting Ebola virus glycoprotein (VRC 608): An open-label phase 1 study. *Lancet* **2019**, *393*, 889–898. [CrossRef]

75. Misasi, J.; Gilman, M.S.A.; Kanekiyo, M.; Gui, M.; Cagigi, A.; Mulangu, S.; Corti, D.; Ledgerwood, J.E.; Lanzavecchia, A.; Cunningham, J.; et al. Structural and molecular basis for Ebola virus neutralization by protective human antibodies. *Science* **2016**, *351*, 1343–1346. [CrossRef] [PubMed]

76. Mulangu, S.; Dodd, L.E.; Davey, R.T.; Mbaya, O.T.; Proschan, M.; Mukadi, D.; Manzo, M.L.; Nzolo, D.; Oloma, A.T.; Ibanda, A.; et al. A Randomized, Controlled Trial of Ebola Virus Disease Therapeutics. *N. Engl. J. Med.* **2019**, *381*, 2293–2303. [CrossRef]

77. Pascal, K.E.; Dudgeon, D.; Trefry, J.C.; Anantpadma, M.; Sakurai, Y.; Murin, C.D.; Turner, H.L.; Fairhurst, J.; Torres, M.; Rafique, A.; et al. Development of Clinical-Stage Human Monoclonal Antibodies That Treat Advanced Ebola Virus Disease in Nonhuman Primates. *J. Infect. Dis.* **2018**, *218*, S612–S626. [CrossRef] [PubMed]

78. Howell, K.; Qiu, X.; Brannan, J.M.; Bryan, C.; Davidson, E.; Holtsberg, F.W.; Wec, A.Z.; Shulenin, S.; Biggins, J.E.; Douglas, R.; et al. Antibody Treatment of Ebola and Sudan Virus Infection via a Uniquely Exposed Epitope within the Glycoprotein Receptor-Binding Site. *Cell Rep.* **2016**, *15*, 1514–1526. [CrossRef]

79. Murin, C.D.; Fusco, M.L.; Bornholdt, Z.; Qiu, X.; Olinger, G.; Zeitlin, L.; Kobinger, G.P.; Ward, A.B.; Saphire, E.O. Structures of protective antibodies reveal sites of vulnerability on Ebola virus. *Proc. Natl. Acad. Sci. USA* **2014**, *111*, 17182–17187. [CrossRef] [PubMed]

80. Davidson, E.; Bryan, C.; Fong, R.H.; Barnes, T.; Pfaff, J.M.; Mabila, M.; Rucker, J.B.; Doranz, B.J. Mechanism of Binding to Ebola Virus Glycoprotein by the ZMapp, ZMAb, and MB-003 Cocktail Antibodies. *J. Virol.* **2015**, *89*, 10982–10992. [CrossRef]

81. Saphire, E.O.; Schendel, S.; Fusco, M.L.; Gangavarapu, K.; Gunn, B.M.; Wec, A.Z.; Halfmann, P.J.; Brannan, J.M.; Herbert, A.S.; Qiu, X.; et al. Systematic Analysis of Monoclonal Antibodies against Ebola Virus GP Defines Features that Contribute to Protection. *Cell* **2018**, *174*, 938–952.e13. [CrossRef] [PubMed]

82. Gilchuk, P.; Murin, C.D.; Milligan, J.C.; Cross, R.W.; Mire, C.E.; Ilinykh, P.A.; Huang, K.; Kuzmina, N.; Altman, P.X.; Hui, S.; et al. Analysis of a Therapeutic Antibody Cocktail Reveals Determinants for Cooperative and Broad Ebolavirus Neutralization. *Immunity* **2020**, *52*, 388–403.e12. [CrossRef]

83. Gilchuk, P.; Kuzmina, N.; Ilinykh, P.A.; Huang, K.; Gunn, B.M.; Bryan, A.; Davidson, E.; Doranz, B.J.; Turner, H.L.; Fusco, M.L.; et al. Multifunctional Pan-ebolavirus Antibody Recognizes a Site of Broad Vulnerability on the Ebolavirus Glycoprotein. *Immunity* **2018**, *49*, 363–374.e10. [CrossRef]

84. Murin, C.D.; Gilchuk, P.; Ilinykh, P.A.; Huang, K.; Kuzmina, N.; Shen, X.; Bruhn, J.F.; Bryan, A.L.; Davidson, E.; Doranz, B.J.; et al. Convergence of a common solution for broad ebolavirus neutralization by glycan cap-directed human antibodies. *Cell Rep.* **2021**, *35*, 108984. [CrossRef]

85. Murin, C.D.; Bruhn, J.F.; Bornholdt, Z.A.; Copps, J.; Stanfield, R.; Ward, A.B. Structural Basis of Pan-Ebolavirus Neutralization by an Antibody Targeting the Glycoprotein Fusion Loop. *Cell Rep.* **2018**, *24*, 2723–2732. [CrossRef]

86. Wec, A.Z.; Bornholdt, Z.A.; He, S.; Herbert, A.S.; Goodwin, E.; Wirchnianski, A.S.; Gunn, B.M.; Zhang, Z.; Zhu, W.; Liu, G.; et al. Development of a Human Antibody Cocktail that Deploys Multiple Functions to Confer Pan-Ebolavirus Protection. *Cell Host Microbe* **2019**, *25*, 39–48.e5. [CrossRef]

87. Brannan, J.M.; He, S.; Howell, K.A.; Prugar, L.I.; Zhu, W.; Vu, H.; Shulenin, S.; Kailasan, S.; Raina, H.; Wong, G.; et al. Post-exposure immunotherapy for two ebolaviruses and Marburg virus in nonhuman primates. *Nat. Commun.* **2019**, *10*, 105. [CrossRef]

88. Mire, C.E.; Geisbert, J.B.; Borisevich, V.; Fenton, K.A.; Agans, K.N.; Flyak, A.I.; Deer, D.J.; Steinkellner, H.; Bohorov, O.; Bohorova, N.; et al. Therapeutic treatment of Marburg and Ravn virus infection in nonhuman primates with a human monoclonal antibody. *Sci. Transl. Med.* **2017**, *9*, eaai8711. [CrossRef]

89. King, L.B.; Fusco, M.L.; Flyak, A.I.; Ilinykh, P.A.; Huang, K.; Gunn, B.; Kirchdoerfer, R.N.; Hastie, K.M.; Sangha, A.K.; Meiler, J.; et al. The Marburgvirus-Neutralizing Human Monoclonal Antibody MR191 Targets a Conserved Site to Block Virus Receptor Binding. *Cell Host Microbe* **2018**, *23*, 101–109.e4. [CrossRef] [PubMed]

90. Wang, H.; Shi, Y.; Song, J.; Qi, J.; Lu, G.; Yan, J.; Gao, G.F. Ebola Viral Glycoprotein Bound to Its Endosomal Receptor Niemann-Pick C1. *Cell* **2016**, *164*, 258–268. [CrossRef] [PubMed]

91. Ridgeback Biotherapeutics. *Drug Label: EBANGA (Ansuvimab-Zykl) for Injection*; Ridgeback Biotherapeutics: Miami, FL, USA, 2020.

92. FDA. *Multi-Discipline Review*; Inmazeb (Atoltivimab, Maftivimab, and Odesivimab-Ebgn) Application Number 761169Orig1s000; FDA: Silver Spring, MD, USA, 2018.

93. Zhao, X.; Howell, K.A.; He, S.; Brannan, J.M.; Wec, A.Z.; Davidson, E.; Turner, H.L.; Chiang, C.-I.; Lei, L.; Fels, J.M.; et al. Immunization-Elicited Broadly Protective Antibody Reveals Ebolavirus Fusion Loop as a Site of Vulnerability. *Cell* **2017**, *169*, 891–904.e15. [CrossRef]

94. Janus, B.M.; Van Dyk, N.; Zhao, X.; Howell, K.A.; Soto, C.; Aman, M.J.; Li, Y.; Fuerst, T.R.; Ofek, G. Structural basis for broad neutralization of ebolaviruses by an antibody targeting the glycoprotein fusion loop. *Nat. Commun.* **2018**, *9*, 3934. [CrossRef] [PubMed]
95. Marano, G.; Vaglio, S.; Pupella, S.; Facco, G.; Catalano, L.; Liumbruno, G.M.; Grazzini, G. Convalescent plasma: New evidence for an old therapeutic tool? *Blood Transfus.* **2016**, *14*, 1–6. [CrossRef]
96. Simon, J. Emil Behring's Medical Culture: From Disinfection to Serotherapy. *Med. Hist.* **2007**, *51*, 201–218. [CrossRef] [PubMed]
97. Dean, C.L.; Hooper, J.W.; Dye, J.M.; Zak, S.E.; Koepsell, S.A.; Corash, L.; Benjamin, R.J.; Kwilas, S.; Bonds, S.; Winkler, A.M.; et al. Characterization of Ebola convalescent plasma donor immune response and psoralen treated plasma in the United States. *Transfusion* **2020**, *60*, 1024–1031. [CrossRef] [PubMed]
98. Delamou, A.; Haba, N.Y.; Saez, A.M.; Gallian, P.; Ronse, M.; Jacobs, J.; Camara, B.S.; Kadio, K.J.-J.O.; Guemou, A.; Kolie, J.P.; et al. Organizing the Donation of Convalescent Plasma for a Therapeutic Clinical Trial on Ebola Virus Disease: The Experience in Guinea. *Am. J. Trop. Med. Hyg.* **2016**, *95*, 647–653. [CrossRef] [PubMed]
99. Van Griensven, J.; Edwards, T.; De Lamballerie, X.; Semple, M.; Gallian, P.; Baize, S.; Horby, P.; Raoul, H.; Magassouba, N.; Antierens, A.; et al. Evaluation of Convalescent Plasma for Ebola Virus Disease in Guinea. *N. Engl. J. Med.* **2016**, *374*, 33–42. [CrossRef]
100. Brown, J.F.; Dye, J.M.; Tozay, S.; Jeh-Mulbah, G.; Wohl, D.A.; Fischer, W.A.; Cunningham, C.K.; Rowe, K.; Zacharias, P.; Van Hasselt, J.; et al. Anti–Ebola Virus Antibody Levels in Convalescent Plasma and Viral Load After Plasma Infusion in Patients with Ebola Virus Disease. *J. Infect. Dis.* **2018**, *218*, 555–562. [CrossRef]
101. Tedder, R.; Samuel, D.; Dicks, S.; Scott, J.T.; Ijaz, S.; Smith, C.C.; Adaken, C.; Cole, C.; Baker, S.; Edwards, T.; et al. Detection, characterization, and enrollment of donors of Ebola convalescent plasma in Sierra Leone. *Transfusion* **2018**, *58*, 1289–1298. [CrossRef]
102. Marzi, A.; Robertson, S.J.; Haddock, E.; Feldmann, F.; Hanley, P.W.; Scott, D.P.; Strong, J.E.; Kobinger, G.; Best, S.M.; Feldmann, H. VSV-EBOV rapidly protects macaques against infection with the 2014/15 Ebola virus outbreak strain. *Science* **2015**, *349*, 739–742. [CrossRef]
103. Jones, S.M.; Ströher, U.; Fernando, L.; Qiu, X.; Alimonti, J.; Melito, P.; Bray, M.; Klenk, H.; Feldmann, H. Assessment of a Vesicular Stomatitis Virus–Based Vaccine by Use of the Mouse Model of Ebola Virus Hemorrhagic Fever. *J. Infect. Dis.* **2007**, *196*, S404–S412. [CrossRef] [PubMed]
104. Marzi, A.; Engelmann, F.; Feldmann, F.; Haberthur, K.; Shupert, W.L.; Brining, D.; Scott, D.P.; Geisbert, T.W.; Kawaoka, Y.; Katze, M.G.; et al. Antibodies are necessary for rVSV/ZEBOV-GP-mediated protection against lethal Ebola virus challenge in nonhuman primates. *Proc. Natl. Acad. Sci. USA* **2013**, *110*, 1893–1898. [CrossRef] [PubMed]
105. Wong, G.; Richardson, J.S.; Pillet, S.; Patel, A.; Qiu, X.; Alimonti, J.; Hogan, J.; Zhang, Y.; Takada, A.; Feldmann, H.; et al. Immune Parameters Correlate with Protection Against Ebola Virus Infection in Rodents and Nonhuman Primates. *Sci. Transl. Med.* **2012**, *4*, 158ra146. [CrossRef]
106. Wong, G.; Audet, J.; Fernando, L.; Fausther-Bovendo, H.; Alimonti, J.B.; Kobinger, G.P.; Qiu, X. Immunization with vesicular stomatitis virus vaccine expressing the Ebola glycoprotein provides sustained long-term protection in rodents. *Vaccine* **2014**, *32*, 5722–5729. [CrossRef]
107. Feldmann, H.; Jones, S.M.; Daddario-DiCaprio, K.M.; Geisbert, J.B.; Ströher, U.; Grolla, A.; Bray, M.; Fritz, E.A.; Fernando, L.; Feldmann, F.; et al. Effective Post-Exposure Treatment of Ebola Infection. *PLoS Pathog.* **2007**, *3*, e2. [CrossRef]
108. Geisbert, T.W.; Daddario-DiCaprio, K.M.; Geisbert, J.B.; Reed, D.; Feldmann, F.; Grolla, A.; Ströher, U.; Fritz, E.A.; Hensley, L.; Jones, S.M.; et al. Vesicular stomatitis virus-based vaccines protect nonhuman primates against aerosol challenge with Ebola and Marburg viruses. *Vaccine* **2008**, *26*, 6894–6900. [CrossRef]
109. Qiu, X.; Fernando, L.; Alimonti, J.B.; Melito, P.L.; Feldmann, F.; Dick, D.; Ströher, U.; Feldmann, H.; Jones, S.M. Mucosal Immunization of Cynomolgus Macaques with the VSVΔG/ZEBOVGP Vaccine Stimulates Strong Ebola GP-Specific Immune Responses. *PLoS ONE* **2009**, *4*, e5547. [CrossRef]
110. Geisbert, T.W.; Daddario-DiCaprio, K.M.; Lewis, M.G.; Geisbert, J.B.; Grolla, A.; Leung, A.; Paragas, J.; Matthias, L.; Smith, M.A.; Jones, S.M.; et al. Vesicular Stomatitis Virus-Based Ebola Vaccine Is Well-Tolerated and Protects Immunocompromised Nonhuman Primates. *PLoS Pathog.* **2008**, *4*, e1000225. [CrossRef] [PubMed]
111. Agnandji, S.T.; Huttner, A.; Zinser, M.E.; Njuguna, P.; Dahlke, C.; Fernandes, J.F.; Yerly, S.; Dayer, J.-A.; Kraehling, V.; Kasonta, R.; et al. Phase 1 Trials of rVSV Ebola Vaccine in Africa and Europe. *N. Engl. J. Med.* **2016**, *374*, 1647–1660. [CrossRef]
112. Regules, J.A.; Beigel, J.; Paolino, K.M.; Voell, J.; Castellano, A.R.; Hu, Z.; Muñoz, P.; Moon, J.E.; Ruck, R.C.; Bennett, J.W.; et al. A Recombinant Vesicular Stomatitis Virus Ebola Vaccine. *N. Engl. J. Med.* **2017**, *376*, 330–341. [CrossRef]
113. Huttner, A.; Dayer, J.-A.; Yerly, S.; Combescure, C.; Auderset, F.; Desmeules, J.; Eickmann, M.; Finckh, A.; Goncalves, A.R.; Hooper, J.; et al. The effect of dose on the safety and immunogenicity of the VSV Ebola candidate vaccine: A randomised double-blind, placebo-controlled phase 1/2 trial. *Lancet Infect. Dis.* **2015**, *15*, 1156–1166. [CrossRef]
114. ElSherif, M.S.; Brown, C.; MacKinnon-Cameron, D.; Li, L.; Racine, T.; Alimonti, J.; Rudge, T.L.; Sabourin, C.; Silvera, P.; Hooper, J.; et al. Assessing the safety and immunogenicity of recombinant vesicular stomatitis virus Ebola vaccine in healthy adults: A randomized clinical trial. *Can. Med. Assoc. J.* **2017**, *189*, E819–E827. [CrossRef]

115. Henao-Restrepo, A.M.; Camacho, A.; Longini, I.M.; Watson, C.; Edmunds, W.J.; Egger, M.; Carroll, M.; Dean, N.E.; Diatta, I.D.; Doumbia, M.; et al. Efficacy and effectiveness of an rVSV-vectored vaccine in preventing Ebola virus disease: Final results from the Guinea ring vaccination, open-label, cluster-randomised trial (Ebola Ça Suffit!). *Lancet* **2017**, *389*, 505–518. [CrossRef]
116. Metzger, W.G.; Vivas-Martínez, S. Questionable efficacy of the rVSV-ZEBOV Ebola vaccine. *Lancet* **2018**, *391*, 1021. [CrossRef]
117. Halperin, S.A.; Das, R.; Onorato, M.T.; Liu, K.; Martin, J.; Grant-Klein, R.J.; Nichols, R.; Coller, B.-A.; Helmond, F.A.; Simon, J.K. Immunogenicity, Lot Consistency, and Extended Safety of rVSVΔG-ZEBOV-GP Vaccine: A Phase 3 Randomized, Double-Blind, Placebo-Controlled Study in Healthy Adults. *J. Infect. Dis.* **2019**, *220*, 1127–1135. [CrossRef] [PubMed]
118. Khurana, S.; Fuentes, S.; Coyle, E.M.; Ravichandran, S.; Davey, R.T., Jr.; Beigel, J. Human antibody repertoire after VSV-Ebola vaccination identifies novel targets and virus-neutralizing IgM antibodies. *Nat. Med.* **2016**, *22*, 1439–1447. [CrossRef]
119. Stephenson, K.E.; Le Gars, M.; Sadoff, J.; de Groot, A.M.; Heerwegh, D.; Truyers, C.; Atyeo, C.; Loos, C.; Chandrashekar, A.; McMahan, K.; et al. Immunogenicity of the Ad26.COV2.S Vaccine for COVID-19. *JAMA* **2021**, *325*, 1535–1544. [CrossRef] [PubMed]
120. Priddy, F.H.; Brown, D.; Kublin, J.; Monahan, K.; Wright, D.P.; Lalezari, J.; Santiago, S.; Marmor, M.; Lally, M.; Novak, R.M.; et al. Safety and Immunogenicity of a Replication-Incompetent Adenovirus Type 5 HIV-1 Clade Bgag/pol/nefVaccine in Healthy Adults. *Clin. Infect. Dis.* **2008**, *46*, 1769–1781. [CrossRef] [PubMed]
121. Cupovic, J.; Ring, S.S.; Onder, L.; Colston, J.M.; Lütge, M.; Cheng, H.-W.; De Martin, A.; Provine, N.M.; Flatz, L.; Oxenius, A.; et al. Adenovirus vector vaccination reprograms pulmonary fibroblastic niches to support protective inflating memory CD8$^+$ T cells. *Nat. Immunol.* **2021**, *22*, 1042–1051. [CrossRef]
122. Chen, T.; Li, D.; Song, Y.; Yang, X.; Liu, Q.; Jin, X.; Zhou, D.; Huang, Z. A heterologous prime-boost Ebola virus vaccine regimen induces durable neutralizing antibody response and prevents Ebola virus-like particle entry in mice. *Antivir. Res.* **2017**, *145*, 54–59. [CrossRef]
123. Pollard, A.J.; Launay, O.; Lelievre, J.-D.; Lacabaratz, C.; Grande, S.; Goldstein, N.; Robinson, C.; Gaddah, A.; Bockstal, V.; Wiedemann, A.; et al. Safety and immunogenicity of a two-dose heterologous Ad26.ZEBOV and MVA-BN-Filo Ebola vaccine regimen in adults in Europe (EBOVAC2): A randomised, observer-blind, participant-blind, placebo-controlled, phase 2 trial. *Lancet Infect. Dis.* **2020**, *21*, 493–506. [CrossRef]
124. Anywaine, Z.; Whitworth, H.; Kaleebu, P.; PrayGod, G.; Shukarev, G.; Manno, D.; Kapiga, S.; Grosskurth, H.; Kalluvya, S.; Bockstal, V.; et al. Safety and Immunogenicity of a 2-Dose Heterologous Vaccination Regimen With Ad26.ZEBOV and MVA-BN-Filo Ebola Vaccines: 12-Month Data From a Phase 1 Randomized Clinical Trial in Uganda and Tanzania. *J. Infect. Dis.* **2019**, *220*, 46–56. [CrossRef] [PubMed]
125. Mutua, G.; Anzala, O.; Luhn, K.; Robinson, C.; Bockstal, V.; Anumendem, D.; Douoguih, M. Safety and Immunogenicity of a 2-Dose Heterologous Vaccine Regimen with Ad26.ZEBOV and MVA-BN-Filo Ebola Vaccines: 12-Month Data from a Phase 1 Randomized Clinical Trial in Nairobi, Kenya. *J. Infect. Dis.* **2019**, *220*, 57–67. [CrossRef]
126. Cross, R.W.; Bornholdt, Z.A.; Prasad, A.N.; Geisbert, J.B.; Borisevich, V.; Agans, K.N.; Deer, D.J.; Melody, K.; Fenton, K.A.; Feldmann, H.; et al. Prior vaccination with rVSV-ZEBOV does not interfere with but improves efficacy of postexposure antibody treatment. *Nat. Commun.* **2020**, *11*, 1–8. [CrossRef] [PubMed]
127. Suder, E.; Furuyama, W.; Feldmann, H.; Marzi, A.; de Wit, E. The vesicular stomatitis virus-based Ebola virus vaccine: From concept to clinical trials. *Hum. Vaccines Immunother.* **2018**, *14*, 2107–2113. [CrossRef]
128. Jones, S.M.; Feldmann, H.; Ströher, U.; Geisbert, J.B.; Fernando, L.; Grolla, A.; Klenk, H.-D.; Sullivan, N.J.; Volchkov, V.; Fritz, E.A.; et al. Live attenuated recombinant vaccine protects nonhuman primates against Ebola and Marburg viruses. *Nat. Med.* **2005**, *11*, 786–790. [CrossRef] [PubMed]
129. Daddario-DiCaprio, K.M.; Geisbert, T.W.; Geisbert, J.B.; Ströher, U.; Hensley, L.E.; Grolla, A.; Fritz, E.A.; Feldmann, F.; Feldmann, H.; Jones, S.M. Cross-Protection against Marburg Virus Strains by Using a Live, Attenuated Recombinant Vaccine. *J. Virol.* **2006**, *80*, 9659–9666. [CrossRef]
130. Mire, C.E.; Geisbert, J.B.; Agans, K.N.; Satterfield, B.; Versteeg, K.M.; Fritz, E.A.; Feldmann, H.; Hensley, L.; Geisbert, T.W. Durability of a Vesicular Stomatitis Virus-Based Marburg Virus Vaccine in Nonhuman Primates. *PLoS ONE* **2014**, *9*, e94355. [CrossRef]
131. Daddario-DiCaprio, K.M.; Geisbert, T.W.; Ströher, U.; Geisbert, J.B.; Grolla, A.; Fritz, E.A.; Fernando, L.; Kagan, E.; Jahrling, P.B.; Hensley, L.; et al. Postexposure protection against Marburg haemorrhagic fever with recombinant vesicular stomatitis virus vectors in non-human primates: An efficacy assessment. *Lancet* **2006**, *367*, 1399–1404. [CrossRef]
132. Geisbert, T.W.; Hensley, L.; Geisbert, J.B.; Leung, A.; Johnson, J.; Grolla, A.; Feldmann, H. Postexposure Treatment of Marburg Virus Infection. *Emerg. Infect. Dis.* **2010**, *16*, 1119–1122. [CrossRef]
133. Shurtleff, A.C.; Biggins, J.E.; Keeney, A.E.; Zumbrun, E.E.; Bloomfield, H.A.; Kuehne, A.; Audet, J.L.; Alfson, K.J.; Griffiths, A.; Olinger, G.G.; et al. Standardization of the Filovirus Plaque Assay for Use in Preclinical Studies. *Viruses* **2012**, *4*, 3511–3530. [CrossRef] [PubMed]
134. HSE. *Biological Agents 137. The Principles, Design and Operation of Containment Level 4 Facilities*; HSE: London, UK, 2006.
135. Steeds, K. Antibody Correlates of Protection for Ebola Virus Infection: Effects of Mutations within the Viral Glycoprotein on Immune Escape. Ph.D. Thesis, University of Liverpool, Liverpool, UK, 2019.

136. Steeds, K.; Hall, Y.; Slack, G.S.; Longet, S.; Strecker, T.; Fehling, S.K.; Wright, E.; Bore, J.A.; Koundouno, F.R.; Konde, M.K.; et al. Pseudotyping of VSV with Ebola virus glycoprotein is superior to HIV-1 for the assessment of neutralising antibodies. *Sci. Rep.* **2020**, *10*, 14289. [CrossRef] [PubMed]

137. Konduru, K.; Shurtleff, A.C.; Bavari, S.; Kaplan, G. High degree of correlation between Ebola virus BSL-4 neutralization assays and pseudotyped VSV BSL-2 fluorescence reduction neutralization test. *J. Virol. Methods* **2018**, *254*, 1–7. [CrossRef] [PubMed]

138. Wilkinson, D.E.; Page, M.; Mattiuzzo, G.; Hassall, M.; Dougall, T.; Rigsby, P.; Stone, L.; Minor, P. Comparison of platform technologies for assaying antibody to Ebola virus. *Vaccine* **2017**, *35*, 1347–1352. [CrossRef]

139. Ilinykh, P.A.; Shen, X.; Flyak, A.; Kuzmina, N.; Ksiazek, T.G.; Crowe, J.E.; Bukreyev, A. Chimeric Filoviruses for Identification and Characterization of Monoclonal Antibodies. *J. Virol.* **2016**, *90*, 3890–3901. [CrossRef]

140. Shedlock, D.J.; Bailey, M.A.; Popernack, P.M.; Cunningham, J.M.; Burton, D.R.; Sullivan, N.J. Antibody-mediated neutralization of Ebola virus can occur by two distinct mechanisms. *Virology* **2010**, *401*, 228–235. [CrossRef]

141. Kaletsky, R.L.; Simmons, G.; Bates, P. Proteolysis of the Ebola Virus Glycoproteins Enhances Virus Binding and Infectivity. *J. Virol.* **2007**, *81*, 13378–13384. [CrossRef]

142. Roguin, L.; Retegui, L.A. Monoclonal Antibodies Inducing Conformational Changes on the Antigen Molecule. *Scand. J. Immunol.* **2003**, *58*, 387–394. [CrossRef] [PubMed]

143. Yu, D.-S.; Weng, T.-H.; Shen, L.; Wu, X.; Hu, C.-Y.; Wang, F.X.; Wu, Z.-G.; Wu, H.-B.; Wu, N.-P.; Li, L.-J.; et al. Development and Characterization of Neutralizing Antibodies Against Zaire Ebolavirus Glycoprotein and Protein 40. *Cell. Physiol. Biochem.* **2018**, *50*, 1055–1067. [CrossRef]

144. Stahelin, R.V. Could the Ebola Virus Matrix Protein VP40 be a Drug Target? *Expert Opin. Ther. Targets* **2013**, *18*, 115–120. [CrossRef]

145. Brioen, P.; Dekegel, D.; Boeyé, A. Neutralization of poliovirus by antibody-mediated polymerization. *Virology* **1983**, *127*, 463–468. [CrossRef]

146. Murin, C.D.; Wilson, I.A.; Ward, A.B. Antibody responses to viral infections: A structural perspective across three different enveloped viruses. *Nat. Microbiol.* **2019**, *4*, 734–747. [CrossRef]

147. Qiao, J.; Li, Y.; Wei, C.; Yang, H.; Yu, J.; Wei, H. Rapid detection of viral antibodies based on multifunctional Staphylococcus aureus nanobioprobes. *Enzym. Microb. Technol.* **2016**, *95*, 94–99. [CrossRef] [PubMed]

148. Pelegrin, M.; Naranjo-Gomez, M.; Piechaczyk, M. Antiviral Monoclonal Antibodies: Can They Be More Than Simple Neutralizing Agents? *Trends Microbiol.* **2015**, *23*, 653–665. [CrossRef]

149. Mellors, J.; Tipton, T.; Longet, S.; Carroll, M. Viral Evasion of the Complement System and Its Importance for Vaccines and Therapeutics. *Front. Immunol.* **2020**, *11*, 1450. [CrossRef]

150. Howell, K.A.; Brannan, J.M.; Bryan, C.; McNeal, A.; Davidson, E.; Turner, H.L.; Vu, H.; Shulenin, S.; He, S.; Kuehne, A.; et al. Cooperativity Enables Non-neutralizing Antibodies to Neutralize Ebolavirus. *Cell Rep.* **2017**, *19*, 413–424. [CrossRef]

151. Kuzmina, N.; Younan, P.; Gilchuk, P.; Santos, R.I.; Flyak, A.I.; Ilinykh, P.A.; Huang, K.; Lubaki, N.M.; Ramanathan, P.; Crowe, J.E.; et al. Antibody-Dependent Enhancement of Ebola Virus Infection by Human Antibodies Isolated from Survivors. *Cell Rep.* **2018**, *24*, 1802–1815.e5. [CrossRef]

152. Weisblum, Y.; Schmidt, F.; Zhang, F.; DaSilva, J.; Poston, D.; Lorenzi, J.C.; Muecksch, F.; Rutkowska, M.; Hoffmann, H.-H.; Michailidis, E.; et al. Escape from neutralizing antibodies by SARS-CoV-2 spike protein variants. *eLife* **2020**, *9*, e61312. [CrossRef]

153. Hiatt, A.; Zeitlin, L.; Whaley, K.J. Multiantibody Strategies for HIV. *Clin. Dev. Immunol.* **2013**, *2013*, 632893. [CrossRef] [PubMed]

154. Yasuhara, A.; Yamayoshi, S.; Ito, M.; Kiso, M.; Yamada, S.; Kawaoka, Y. Isolation and Characterization of Human Monoclonal Antibodies That Recognize the Influenza A(H1N1)pdm09 Virus Hemagglutinin Receptor-Binding Site and Rarely Yield Escape Mutant Viruses. *Front. Microbiol.* **2018**, *9*, 2660. [CrossRef]

155. Taylor, P.C.; Adams, A.C.; Hufford, M.M.; de la Torre, I.; Winthrop, K.; Gottlieb, R.L. Neutralizing monoclonal antibodies for treatment of COVID-19. *Nat. Rev. Immunol.* **2021**, *21*, 382–393. [CrossRef] [PubMed]

156. Bramble, M.S.; Hoff, N.; Gilchuk, P.; Mukadi, P.; Lu, K.; Doshi, R.H.; Steffen, I.; Nicholson, B.P.; Lipson, A.; Vashist, N.; et al. Pan-Filovirus Serum Neutralizing Antibodies in a Subset of Congolese Ebolavirus Infection Survivors. *J. Infect. Dis.* **2018**, *218*, 1929–1936. [CrossRef] [PubMed]

157. Keck, Z.-Y.; Enterlein, S.G.; Howell, K.A.; Vu, H.; Shulenin, S.; Warfield, K.L.; Froude, J.W.; Araghi, N.; Douglas, R.; Biggins, J.; et al. Macaque Monoclonal Antibodies Targeting Novel Conserved Epitopes within Filovirus Glycoprotein. *J. Virol.* **2016**, *90*, 279–291. [CrossRef] [PubMed]

Review

Development and Structural Analysis of Antibody Therapeutics for Filoviruses

Xiaoying Yu [1] and Erica Ollmann Saphire [1,2,*]

1   Center for Infectious Disease and Vaccine Discovery, La Jolla Institute for Immunology, 9420 Athena Circle, La Jolla, CA 92037, USA; xiaoyingyu@lji.org
2   Department of Medicine, University of California, San Diego, La Jolla, CA 92093, USA
*   Correspondence: erica@lji.org; Tel.: +1-858-752-6791

Abstract: The filoviruses, including ebolaviruses and marburgviruses, are among the world's deadliest pathogens. As the only surface-exposed protein on mature virions, their glycoprotein GP is the focus of current therapeutic monoclonal antibody discovery efforts. With recent technological developments, potent antibodies have been identified from immunized animals and human survivors of virus infections and have been characterized functionally and structurally. Structural insight into how the most successful antibodies target GP further guides vaccine development. Here we review the recent developments in the identification and characterization of neutralizing antibodies and cocktail immunotherapies.

Keywords: antibody therapeutic; cryo-EM; ebolavirus; filovirus; marburgvirus; structural biology; X-ray crystallography

Citation: Yu, X.; Saphire, E.O. Development and Structural Analysis of Antibody Therapeutics for Filoviruses. *Pathogens* **2022**, *11*, 374. https://doi.org/10.3390/pathogens11030374

Academic Editors: Philipp A. Ilinykh and Kai Huang

Received: 22 February 2022
Accepted: 17 March 2022
Published: 18 March 2022

1. Introduction

The Filoviruses belong to the family *Filoviridae* and are among the world's deadliest pathogens. Among the six genera are the *Ebolaviruses*, the *Marburgviruses*, the *Cuevavirus*, the recently discovered *Dianlovi* rus [1], *Striavirus* [2], and the *Thamnovirus* [2]. There are 11 species in total, of which the ebolaviruses and marburgviruses are known to cause severe disease in humans. The genus *ebolavirus* includes six known viruses that are each antigenically distinct and named after the location of the disease outbreak where they were first identified. These include Ebola virus (EBOV), Sudan virus (SUDV), Bundibugyo virus (BDBV), Reston virus (RESTV), Taï Forest virus (TAFV), and Bombali virus (BOMV). The *Marburgvirus* genus contains Marburg virus (MARV) and its variant Ravn (RAVV).

The first filovirus to be identified, MARV, was discovered in 1967 when several laboratory workers in Germany developed hemorrhagic fever after handling tissues from non-human primates (NHPs). A total of 31 people were infected, and 7 died [3]. EBOV was first identified in 1976 when two separate outbreaks occurred in northern Zaire (now the Democratic Republic of Congo, DRC) and southern Sudan. Each outbreak resulted in hundreds of cases with 88% and 53% case fatality, respectively [4,5]. *Ebolavirus* has since appeared sporadically in Africa. The largest outbreak to date, which occurred in West Africa from 2014 to 2016, caused more than 28,600 infections and more than 11,300 deaths from Ebola Virus Disease (EVD). In 2021, two additional outbreaks occurred in the Democratic Republic of Congo [6] and Guinea [7]. Other filoviruses have similar outbreak potential and lethality. The most recent significant emergence of MARV in Angola had a case fatality rate of 90% [8]. Meanwhile, Sudan virus (SUDV) and the newly emergent Bundibugyo virus (BDBV) have case fatality rates of ~50% and 25–50% [9], respectively.

Symptoms of EVD include fever, headache, muscle pain, weakness, fatigue, diarrhea, vomiting, stomach pain, and hemorrhage (severe bleeding) [10,11]. The infection prodrome of filoviruses is virtually identical to common, co-circulating diseases like typhoid fever and malaria [10]. As such, early diagnosis, particularly for those cases early in a disease outbreak, is challenging. Given this potential delay in diagnosis, the therapeutic window for potential treatments must be broad so that treatments are effective even if delivered late in a disease course. Traditional approaches involving post-exposure vaccination and small molecule interventions required almost immediate administration to be effective [12]. Currently, monoclonal antibody (mAb) therapy has been shown to be the most effective route of therapy after symptoms appear, and can confer 100% protection for non-human primates (NHPs), even if administered as late as 5 days post-challenge [13,14]. Therefore, studies of antibodies against filoviruses are an important source for potential reliable therapeutics.

In recent years, progress has been made towards vaccine and treatment development. The first vaccine to be approved, Ervebo, is rVSV-based and was tested in an open-label, cluster-randomized ring vaccination trial in Guinea in 2015 [15], deployed in 2018 in the DRC under compassionate use, before gaining approval from the United States Food and Drug Administration (FDA) in late 2019 [16,17]. Another vaccine candidate utilizes a two-dose heterologous vaccination regimen with a replication-deficient adenovirus type 26 vector-based vaccine expressing a Zaire Ebola virus glycoprotein (Ad26.ZEBOV) and a modified vaccinia Ankara (MVA) vector-based vaccine, encoding glycoproteins from the Zaire EBOV, SUDV, and MARV as well as TAFV nucleoprotein (MVA-BN-Filo). This vaccine has been granted marketing authorization by the European Medicines Agency in the European Union [18,19].

The mAb monotherapy mAb114 and antibody cocktail REGN-EB3 were tested in clinical trials and proved to be effective against EBOV; both were granted FDA approval to treat EVD, and showed superior outcomes in reducing mortality compared to ZMapp and remdesivir [20–22]. The longer and multiple-dose regimen required for ZMapp and remdesivir administration could contribute to the slower rate of viral clearance of patients in those groups, and further lead to the difference in mortality between groups [20]. The intrinsic difference between the patient conditions among the four groups may contribute to the variation in treatment protection outcomes [20]. Notably, however, the therapeutic antibodies for humans approved thus far are only effective against EBOV. None show broad reactivity against other pathogenic filoviruses. Efficacious treatments against a range of pathogenic filoviruses are urgently needed.

2. Viral Entry and Glycoprotein Structure

Filoviruses are enveloped, non-segmented, negative-sense RNA viruses that have a characteristic filamentous shape. The genome has seven genes that encode eight proteins (Figure 1). Six proteins are encoded by the corresponding viral genes, including VP24, NP, VP30, VP35, VP40 (matrix protein), and the RNA-dependent RNA polymerase (L). The ebolavirus *GP* gene expresses two major products: the trimeric glycoprotein, termed GP, which is displayed on the viral surface, and a dimeric, soluble version, termed sGP that represents the majority (80%) of the transcripts and is abundantly shed from infected cells [23–26]. GP and sGP share 295 amino acids and have some similarities in the folding of the monomeric unit [23–26]. The function of sGP remains unclear [23,27], but it has been proposed to act as an immune decoy [28]. Indeed, multiple antibodies cross-react with sGP and GP [29,30]. These antibodies may be absorbed by the much more abundant sGP and thus unavailable to neutralize virtual-surface GP. Many cross-reactive antibodies have a higher affinity for sGP [29,31], and its abundance indicates it may be a major antigen in a natural infection. Interestingly, marburgviruses and dianloviruses do not produce sGP; cuevaviruses express sGP similarly to ebolaviruses [31].

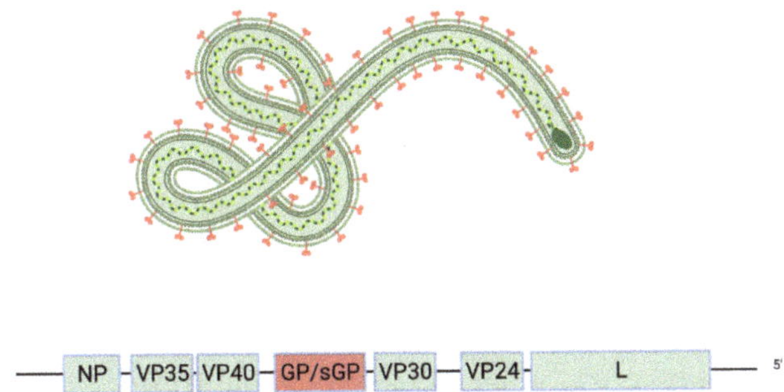

Figure 1. Schematic of the Ebola virus genome and virion. The glycoprotein GP (red) is the only viral protein displayed on the virion surface.

GP mediates attachment and entry into target cells. As the only surface-exposed protein on mature virions, GP is the focus of current filovirus therapeutic mAb discovery efforts (Figure 2) [32]. Filovirus GPs share a common core fold and trimeric organization, but are antigenically distinct due to species-specific sequence differences of up to 70%. The ebolavirus GP monomer comprises GP1 and GP2 subunits that are anchored together by a single GP1-GP2 disulfide bond [33]. The larger GP1 subunit harbors the receptor-binding site (RBS), the glycan cap domain, and the heavily glycosylated mucin-like domain (MLD). GP2 contains the membrane fusion machinery, including the internal fusion loop (IFL), two heptad repeats (HR1 and HR2), the membrane-proximal external region (MPER), and the transmembrane domain (TM) [25]. The GP2 subunit, particularly the IFL and stalk regions, has greater sequence conservation than GP1 among filoviruses. Marburgvirus GPs have a similar arrangement (Figure 2C,D). However, in marburgvirus GP, the furin cleavage site is shifted towards the N-terminus (residue 435 for Marburgvirus vs. 501 for ebolavirus), the region corresponding to the MLD is split into two halves: the major portion of the MLD is attached to the C terminus of GP1. The minor portion (residues 436–501) is attached to the N terminus of the GP2 and is termed the wing domain [34,35].

Filoviruses enter cells via macropinocytosis [36–38]. Once in the endosome, GP is cleaved by host endosomal cathepsins that remove both the MLD and the glycan cap [39–41] to form cleaved GP (GP$_{CL}$). In GP$_{CL}$, the RBS for the cellular receptor Niemann-Pick C1 (NPC1) is exposed at the apex of GP1 [42]. Following receptor binding, the fusion subunit in GP$_{CL}$ rearranges into a six-helix bundle that mediates fusion between host and virus membranes [43].

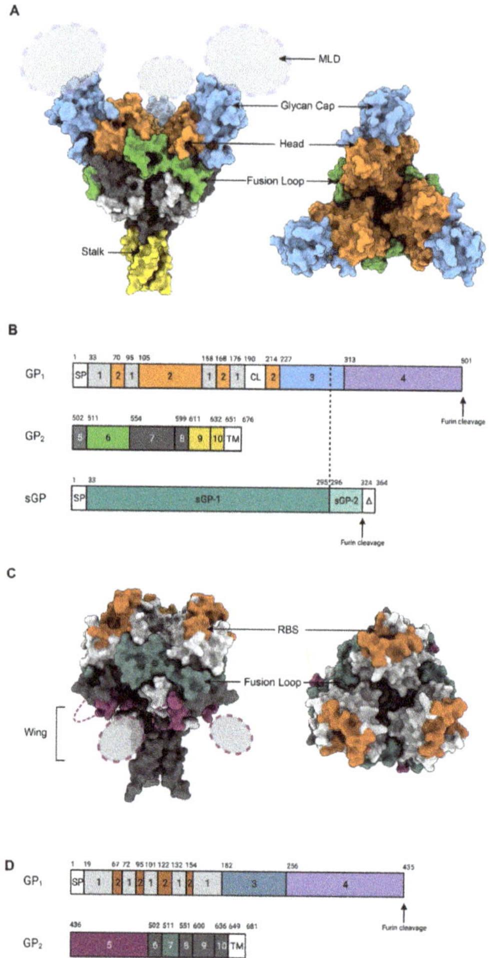

Figure 2. Epitopes on the GP surface. (**A**) Surface representation of Ebola Virus GP structure (PDB: 5JQ3) colored by domain. Side view and top view of Ebola virus GP are illustrated. (**B**). Schematic of the EBOV GP sequence. Amino acid numbering is at top, and polypeptide regions that form key domains are numbered in the center of the schematic blocks. 1: portions of the N-terminus of GP1 that form the base, 2: receptor-binding head, 3: glycan cap, 4: mucin-like domain (MLD), 5: GP2 N-terminal peptide; 6: fusion loop, 7: Heptad repeat 1 (HR1), 8 and 9: Heptad repeat 2 (HR2); 9: stalk; and 10: and membrane-proximal external region (MPER), respectively. Other regions include SP: signal peptide, and TM: transmembrane domain. The organization of sGP is illustrated below. The first 295 residues are identical to those in GP1 (labelled sGP-1). Residues 296 through 324 are unique to sGP (labelled sGP-2). The C-terminal sequence, termed delta peptide, is released from sGP by furin cleavage. (**C**) The surface representation of Marburg Virus GP structure (PDB: 6BP2) colored by domain. Side view and top view of Marburg virus GP are illustrated. (**D**) Schematic of the MARV GP sequence. Amino acid numbering is at top. 1–2: GP1, with 2 for RBS; 3: glycan cap, 4: MLD, 5: wing; 6: N-terminal loop: 7: fusion loop, 8: HR1, 9: HR2; 10: MPER; SP: signal peptide, and TM: transmembrane domain.

3. Efforts for Antibody Discovery

Analyses of antibody responses in human survivors of virus infection can outline key characteristics of antibodies elicited in response to infection and provide an important basis for the development of therapeutic antibodies.

Beginning in the early 1990s, isolation of neutralizing mAbs was mostly enabled by display library approaches, such as phage display [44,45]. Early antibodies like KZ52 and others were discovered from human survivors of the 1995 Kikwit outbreak in DRC [46]. These antibodies facilitated and enhanced understanding of the virus and permitted the determination of the structure of EBOV GP in its pre-fusion conformation [25]. Screening of antibody-secreting hybridomas from immunized mice produced a panel of novel antibodies that could be categorized into multiple different epitope groups [47]. This classification guided the formation of several successful mAb cocktails, including MB-003, ZMAb, and ZMapp, which had non-overlapping binding sites and high efficacy in NHP studies of EBOV infection [13,48,49].

Technological advancements, such as single B cell isolation and next-generation sequencing, accelerated large-scale mAb discovery, direct functional analysis, and exploration of the Ab maturation process. Potent antibodies were subsequently discovered in immunized animals [50–52], human survivors [30,53–55], and vaccinated human volunteers in clinical trials [56,57]. Notably, antibodies in these panels target a spectrum of epitopes on GP, and several neutralize broadly are active against several filovirus species. Several therapeutic antibody cocktails, including REGN-EB3, FVM04/CA45, MBP134^{AF}, rEBOV-520/548, rEBOV-442/515, and 1C3/1C11, were generated and shown to be highly effective [52,58–62].

The functional activity of antibodies isolated during discovery efforts can be evaluated by in vitro neutralization assays to determine whether they block infection by one or more ebolaviruses together with structural biology to reveal the molecular basis for protective activity. Neutralization can be analyzed using authentic virus under BSL-4 containment [63] or at lower biosafety levels using model systems. The biologically contained Ebola virus ΔVP30 system can be performed at BSL-3 [64]. In this system, the entire open reading frame of VP30, which is an essential transcription factor for EBOV replication, is deleted to generate a replication-deficient particle. The VP30 needed for replication is supplied in *trans* through Vero cells that stably express VP30 such that EBOVΔVP30 can undergo multiple replication cycles only in VeroVP30 cells, but not the parental cells that lack VP30. In this system, all viral antigens and proteins, including sGP, are stably expressed. A second neutralization system involves recombinant vesicular stomatitis virus (VSV) bearing EBOV GP [65] and can be performed at BSL-2. This method uses a recombinant VSV (rVSV) in which the VSV-G gene is replaced with filovirus GP that is then displayed on the rhabdoviral surface [65–68]. These pseudovirus constructs typically carry a fluorescence reporter such as GFP to monitor infection, and are frequently modified so that only GP, and not secreted sGP, is expressed.

A systematic analysis of 171 mAbs by The Viral Hemorrhagic Fever Immunotherapeutics Consortium (VIC) compared the readouts of 3 different neutralization assays by epitope and level of in vivo protection [31]. There were several differences in the performance of the 171 mAbs across the 3 assays. For example, the authentic EBOV assay was more forgiving: a group of glycan-cap-directed antibodies only neutralized authentic EBOV and no model system. The ΔVP30 system was more stringent: fewer antibodies demonstrated neutralization overall. The results that correlated best with in vivo protection, however, were those assays that contained sGP (i.e., authentic EBOV and ΔVP30), as well as the fraction left un-neutralized, which was measured in rVSV assay. Antibodies that failed to completely neutralize rVSV at the highest concentration similarly failed to protect in the mouse model.

A small fraction of antibodies had no neutralization activity in any assay, yet nevertheless protected in a mouse model of EBOV infection. In high-throughput systems-serology assays, which examined the contribution of immune effector functions like phagocytosis and activation of natural killer (NK) cells to in vivo antibody efficacy, these antibodies had high immune effector function activity, indicating the potential contribution of Fc-mediated activity to protection [31,69]. By comparison of the in vivo activity of the Fc variants of mAbs from different epitope groups, a previous study revealed the differential requirements for FcγR to achieve protection [70]. MAbs targeting the membrane-proximal regions (HR2 and MPER) or the MLD do not inspire Fc–FcγR interactions. However, the mAbs that contact to chalice bowl region or the fusion loop are linked to FcγR engagement. Moreover, by applying an Fc engineering platform, a library of Fc variants of a specific mAb could be generated to compare how Fc effector functions correlate with mAb protection performance. Based on the results of functional characterization of the variants and the in vivo protection in mice, Fc variants with high complement activity, yet moderate and balanced ADCC activity are more protective against viral infection in vivo [71]. Together, the previous results suggested that effective antibody treatments should comprise mAbs that achieve both neutralization and immune effector functions. The VIC study and other work [31,51,53,54] also revealed the epitopes at which broader ebolavirus cross-reactivity can be achieved.

Structural biology has served a vital role in increasing our understanding of the molecular mechanisms underlying antibody-mediated neutralization of many viral infections. In particular, atomic resolution structures obtained using cryogenic electron microscopy (cryo-EM) or X-ray crystallography allow exploration of the fine details of binding between antibodies and virus proteins like GP to reveal crucial information about molecular interactions and the basis of cross-reactivity. Combining structural studies with virology and in vitro biochemistry allows a thorough evaluation of therapeutic candidates and their target interactions.

3.1. Structural Biology to Reveal Epitopes on GP Targeted by Antibodies

Antibodies isolated from immunized animals or patients infected with ebolavirus have been shown to target several different regions on the surface of the GP trimer. Epitope mapping can be achieved rapidly using competition binding assays or negative stain electron microscopy (EM). However, to definitively understand the antibody interactions with glycoprotein, a high-resolution structure by X-ray crystallography or cryo-EM is required. Structural characterizations of these antibodies in complex with full-length GP ectodomain or GP peptides provide detailed information of the neutralizing mechanism and inform new approaches for broad immunotherapy and vaccine design. The recent development of single-particle cryo-EM capabilities facilitated the determination of more structures of antibody-glycoprotein complexes, including those that involve asymmetric interactions that are difficult to assess by crystallography. As a whole, the structural analyses illustrate how antibodies target the various epitopes on the GP surface, particularly those epitopes in highly conserved regions, to achieve high potency and/or cross-reactivity.

3.2. mAbs Targeting the Glycan Cap

In EBOV, the glycan cap spans between residues 227 and 312, and the majority of residues (amino acids 227–295) are present in both GP and sGP (Figure 2). Thus, antibodies targeting the GP glycan cap typically also react with the abundant, non-structural sGP. If these antibodies were elicited by natural infection, they may, in fact, have been elicited against sGP, which is at least five-fold more abundant than membrane-bound GP. One component of the therapeutic cocktail ZMapp [13,72,73], 13C6 [74], targets the glycan cap and offers protection in in vivo models of infection despite having low neutralizing potency [29,72].

Antibodies against the glycan cap, including 13C6, can be characterized by higher levels of immune effector functions [69]. Some neutralize as well, and several have been characterized functionally and structurally, such as EBOV-548 and EBOV-296 (Figure 3A) [60,75]. Approaching the glycan cap via different angles, anti-glycan cap mAbs with GP largely involve CDRH3 or CDRH2 that mimic and displace the β18-18′ hairpin, which acts as an anchor for the MLD. The proposed mechanism of neutralization for these anti-glycan cap antibodies is blockage of the cathepsin cleavage event that is required for RBS exposure and viral entry [75]. Higher numbers of contacts between a mAb and the MLD cradle are shown to introduce instability in the GP trimer, and thus these antibodies can synergize with those that target the fusion loop. Other glycan-cap targeting mAbs include the Q206, Q314, and Q411 antibodies identified in immunized macaques, which provide partial protection in a mouse model of EBOV challenge [76]. Overall, mAbs in this group are usually potent but rarely have broad neutralizing activity. However, a combination of both neutralizing and effector functions and the demonstrated synergistic effect when pairing with mAbs that target the fusion loop, make some of the more potent glycan cap mAbs good candidates for inclusion in therapeutic cocktails [54,60,75].

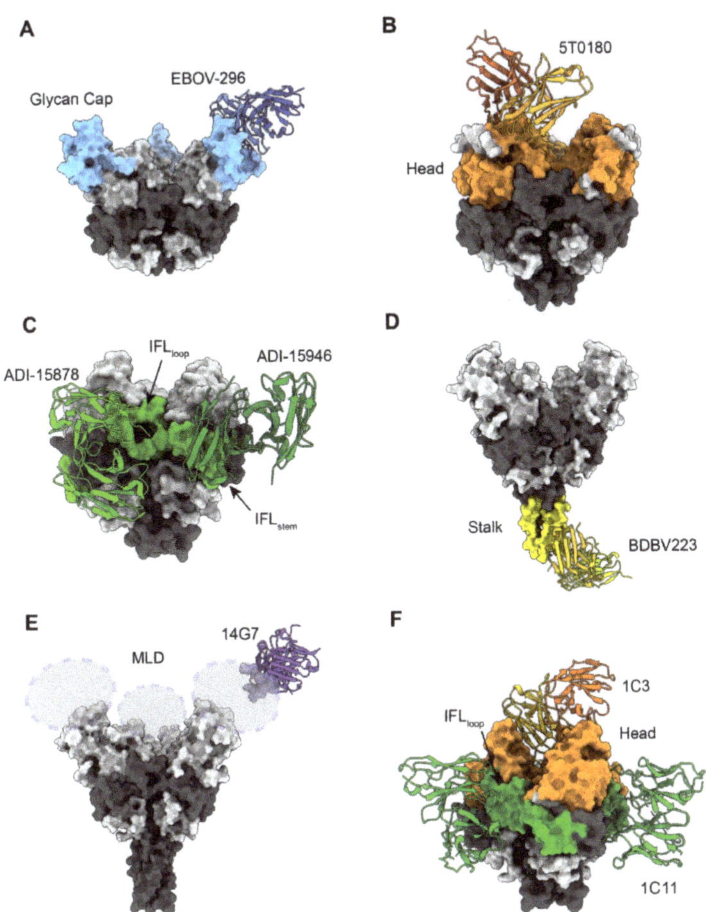

Figure 3. Structural models of neutralizing antibody recognition against key epitopes on the GP surface. GP epitopes are colored as in Figure 2. The variable regions of the antibodies targeting different epitopes are shown in cartoon representation. (**A**) Glycan cap-targeting antibody EBOV-296 (PDB: 7KF9) (**B**) Head-region targeting antibody 5T0180 (PDB:6S8J) (**C**) IFL-targeting antibody ADI-15878 and ADI-15946 (PDB: 6EA5, 6MAM) (**D**) Stalk region-targeting antibody BDBV 223 (PDB: 6N7J, 5JQ3) (**E**) MLD-targeting antibody 14G7 (PDB: 2Y6S. 5JQ3) (**F**) An example of a broad neutralizing antibody cocktail 1C3 and 1C11, with 2 antibodies targeting the head and IFL, respectively. (PDB: 7SWD).

3.3. mAbs Targeting the Apex/Head/Receptor Binding Region of GP

The GP1 Head epitope lies under the glycan cap and contains residues that are part of the RBS. In the late endosome, primed GP_{CL} exhibits a fully exposed RBS that is competent for binding to domain C of NPC1 (NPC1-C) [42]. In contrast to ebolaviruses, the glycan cap of marburgviruses provides a less complete shield and the RBS is more exposed prior to cleavage, such that several antibodies including MR78 and MR191 can target the RBS directly [77–79]. MR78 and MR191 somewhat mimic interactions made by a loop of NPC1-C, which contains aromatic residues that can reach into a hydrophobic cavity on GP_{CL} [42].

The monotherapy mAb114, which was isolated from a survivor of the 1995 Kikwit EVD outbreak, and is approved to treat EVD, targets the head epitope in the RBS via a near-vertical angle to block receptor binding [80]. FVM04, isolated from immunized macaques [50], binds to the inner chalice of the GP trimer near the glycan cap with a tilted angle, so that from low-resolution negative stain EM, only one Fab bound to GP trimer can be visualized [81]. FVM04 can bind and neutralize both EBOV and SUDV but has lower activity toward BDBV [81]. Other head-targeting antibodies (5T0180, 1T0227, and 3T0265), isolated from rVSV-ZEBOV vaccine recipients, also bind the RBS and block NPC1-C, while avoiding contact with the glycan cap region [82] (Figure 3B). Although the residues within the footprint of these 3 antibodies are mostly conserved, a key difference between EBOV and BDBV/SUDV at residue 224 (G in EBOV, N in BDBV/SUDV) sterically prevents their binding to BDBV/SUDV GP, thus limiting the breadth of antibody potency [82]. Overall, previously discovered head targeting neutralizing antibodies are potently neutralizing and protective, but limited in breadth.

A recently published apex-targeting antibody 1C3 is unique among currently characterized EBOV antibodies in that it targets the center of the GP chalice and binds one Fab to one GP trimer (Figure 3F) [62]. 1C3 potently neutralizes both EBOV and SUDV. Although 1C3 does not neutralize BDBV, it does bind to recombinant BDBV GP in ELISA, suggesting that this antibody could potentially contribute to protection against BDBV infection through Fc-dependent effector functions [62]. The asymmetric binding with 1:1 stoichiometry of 1C3 allows more variability within its footprint [83]. Notably, the quaternary recognition of 1C3 is specific for GP and not shed sGP, which may provide an advantage for mAb therapeutic candidates.

3.4. mAbs Targeting Internal Fusion Loop (IFL)

The IFL region (residues 511–554) plays a critical role in viral-host membrane fusion. This important role translates to a high degree of sequence conservation (60–70%), which makes the IFL an ideal epitope for cross-reactive antibodies. The IFL region can be further divided into stem/base (IFL_{stem}; residues 511–520 and 543–554) and the loop/paddle (IFL_{loop}; residues 521–542) [84]. MAbs targeting the IFL usually also contact parts of the base epitope, which is immediately adjacent to the IFL and forms the base of the "bowl" of the GP chalice. The central span of GP, including both the IFL and base, has been termed the "waist" and is targeted by antibodies via a continuum of antibody epitopes [85]. Several cross-reactive antibodies identified thus far have been categorized into the IFL targeting group, including CA45 [86,87], 6D6 [88], ADI-15946 [89], ADI-15878 [85], 2G1 [90], EBOV-520 [60], EBOV-515 [61], and 1C3 [62]. Among these mAbs, 6D6, ADI-15878, and 1C3 have similar footprints that overlap and include the IFL_{loop} region and part of the base region (although the 6D6 complex structure has been resolved only at low-resolution with negative stain EM). Meanwhile, CA45, ADI-15946, EBOV-520, and EBOV-515 have similar footprints that include both the IFL_{stem} and other parts of the base region (Figure 3C). 2G1, isolated from a vaccinated donor and which cross-neutralizes pseudotyped EBOV, SUDV, and BDBV, was determined to bind GP2 by competition assays and was further mapped to the fusion loop by computational modeling [90].

ADI-15878 contacts the IFL_{loop} and the portion of the base termed the N-terminal pocket, which is occupied by the flexible N-terminal tail of GP2 in the GP apo-structure.

The residues that line the N-terminal pocket are highly conserved, but those in the flexible N-terminal tail are not. Therefore, the ability of the ADI-15878 CDRs to reach the conserved pocket region underneath the N-terminal tail provides the cross-reactivity against different ebolaviruses [85]. CA45 targets both the IFL_{stem} and a region termed the DFF cavity [87,91], which is occupied by a short flexible loop that includes residues 192–194 near the cathepsin cleavage site (residues DFF) in the apo GP structure. The footprint of 1C11 partially overlaps that of ADI-15878, but is shifted upwards [62].

ADI-15946, EBOV-520, and EBOV-515 all contact the IFL_{stem} and a region termed the 3_{10} pocket, the core of which encompasses residues 71–75 of GP1. The pocket extends to surrounding residues, including 76–78 of GP1 and 510–516 of GP2 [60,89]. This region is occupied by the β17-β18 loop (residues 287–291), which is part of the glycan cap in the uncleaved GP apo-structure and is exposed in GP_{CL} following removal of the glycan cap. Similar to the crystal structure of the ADI-15946-GP_{CL} complex, the cryo-EM structure of EBOV-520 in complex with uncleaved GP ectodomain shows that the CDRH3 loop of the antibody contacts the 3_{10} pocket [60]. Although the glycan cap region is intact in the EBOV-520 complex structure, the β17-β18 loop cannot be visualized, suggesting that this flexible loop is displaced upon contact with the mAb [60].

ADI-15946 potently neutralizes EBOV and BDBV, but not SUDV, whereas EBOV-520 neutralizes all three viruses. Comparing the footprints of the two mAbs, EBOV-520 is shifted slightly upward to avoid non-conserved position 506 (N506 in EBOV and R506 in SUDV), which may allow the extra reactivity towards SUDV GP [60]. The study on ADI-15946 also provided a good example of structure-based rational engineering. Three residues were substituted to reduce steric and charge clashes with the non-conserved residues in SUDV (R100A in CDRH3) or to improve binding by generating a double tyrosine binding motif (S65Y and F67Y in the FRL3). The designed mAb variant successfully expanded the breadth of ADI-15946 to enhance its binding and neutralization against SUDV [89].

Overall, mAbs that target the IFL region enlist a variety of approaches to contact the most conserved region on the GP surface and showcase the largest number of broadly neutralizing antibodies that have been discovered and characterized to date. The identification of the flexible loops in GP that potentially compete with the mAbs from this group suggests that removal of such regions would contribute to improved antigens and facilitate the development of more antibodies that target these ideal epitopes.

3.5. mAbs Targeting GP Stalk and MPER Region

The stalk/MPER region lies near the C terminus of GP2 above the transmembrane domain. As part of the fusion machinery, the stalk/MPER region has high sequence conservation across filovirus species and is a prime target for mAbs that have broad potency. Negative stain EM was used to map the binding footprint of several cross-reactive antibodies that target the stalk region (BDBV 223, BDBV 317, BDBV 340, and ADI-16061) [54,55]. A high-resolution X-ray crystal structure of the Fab from BDBV 223, isolated from a survivor of BDBV infection, was determined in complex with a synthetic peptide of the epitope region was determined [92] (Figure 3D). Interestingly, the alignment of the complex structure to the GP trimer structure and tomographic reconstruction of the GP trimer on the virus membrane [93] revealed that BDBV 223 binding interferes with the trimeric bundle assembly and anchoring of the GP spike in the viral membrane. Thus, interference with the six-helix bundle formation needed to drive membrane fusion could be a key mechanism by which BDBV 223 neutralizes infection [92].

3.6. mAbs Targeting Mucin-Like Domain

The mucin-like domain (MLD) is a heavily glycosylated region located on the C-terminus of GP1 that shields the top of the GP trimer. The MLD has the most sequence variation among various species. Due to its highly flexible nature, the structure of MLD is not well characterized, and thus the mAbs targeting MLD are less understood. The residues required for binding of MLD mAbs were identified by peptide binding assays [47] or by alanine scanning, which evaluates how mutations at individual residues affect binding to EBOV GP [74]. Several mAbs targeting EBOV MLD have been discovered, including the MB-003 cocktail mAbs 6D8 and 13F6, which exhibit low or no neutralization in vitro [47], yet protected in the mouse challenge model, and, collectively with 13C6, provide protection in the NHP model [49]. Efforts have been made to co-crystalize mAb Fab fragments in complex with peptides of identified epitopes. Such structures have been determined for 13F6 [94] and another mAb targeting the MLD, 14G7 [95] (Figure 3E).

In marburgviruses, the unique wing domain located at the N terminus of GP2 is also part of the MLD. Although this domain has not been fully characterized structurally and functionally, four mAbs targeting this region have shown 90–100% protection in the mouse challenge model [35]. Studies on two wing-specific antibodies, MR228 and MR235, identified from human survivors revealed more features of mAbs in this epitope group. MR228 is non-neutralizing but protective in the mouse model, and its protective activity is likely mediated by Fc effector functions, specifically the engagement of FcγRs [96]. MR235 does not protect in in vivo models of infections, yet cooperatively enhances binding of RBS-targeting neutralizing antibodies by facilitating the structural rearrangement of marburgvirus GP [96]. Overall, mAbs targeting the MLD are less likely to be neutralizing [31] but may offer protection through Fc-mediated functions [74,96].

3.7. mAb Cocktail Immunotherapies

For more than a decade, studies exploring mAbs against filoviruses have demonstrated the potential of using a single or a cocktail of mAbs as immunotherapy, leading to two approved mAb treatments, mAb114 and the cocktail REGN-EB3 in 2020. However, the uncertainty of the causative agents of the next viral outbreak requires a versatile toolbox. The usage of high quantities of mAbs to treat disease caused by filovirus infection presents challenges in production and in cost. Therefore, the development of mAb cocktails as immunotherapies aims to achieve broader reactivity and lower dosage.

First-generation cocktail immunotherapies such as MB-003, ZMAb, and ZMapp can protect against EBOV challenge in NHP [13,48,49]. MB-003 contains antibodies against the MLD and the glycan cap, whereas ZMAb is composed of antibodies against the glycan cap and the base domain. ZMapp is derived from both MB-003 and ZMAb, and combines one mAb from MB-003 with two mAbs from ZMAb, with one mAb (13C6) against the glycan cap and two against the base (2G4 and 4G7) [72,74]. ZMapp was the first antibody cocktail shown to reverse severe disease in the NHP model [13]. During the 2014–2016 outbreak in West Africa, the ZMapp and ZMAb cocktails were used to treat 25 EVD patients under compassionate use protocols in several countries [97]. However, the benefits of the cocktail therapeutics themselves could not be determined definitively since these patients also received other aggressive supportive measures [98]. In a randomized controlled clinical trial, administration of ZMapp was beneficial against human EVD but did not meet the efficacy threshold compared to patients who received the current standard of care alone [99].

REGN-EB3 is a second-generation cocktail of three mAbs, REGN3470, REGN3471, and REGN3479, each isolated from Velocimmune mice, which have human immunoglobulin variable regions [52]. REGN3470 targets the glycan cap from the side, with a binding angle parallel to the viral surface. REGN3471 targets the GP1 head region at the inner chalice, with an angle perpendicular to the viral surface, and REGN3479 targets the fusion loop [52]. REGN-EB3 was superior to ZMapp in reducing EVD mortality in a randomized clinical trial [20], and was approved by the FDA in 2020. An antibody against the head domain, mAb114 [14], was similarly effective as a monotherapy [20]. A two-antibody cocktail including rEBOV-520 that targets the fusion loop region/base area, and rEBOV-548, which targets the glycan cap, is also effective in protecting NHP from EBOV infection [60].

With the uncertainty of viral species responsible for the next outbreak, the next generation of immunotherapy ideally will offer a cross-protective cocktail. In recent years, several mAb cocktails have been characterized and investigated in NHPs to demonstrate protection against viral infection. Two broadly neutralizing mAbs, FVM04 and CA45, have been evaluated as a cocktail in NHPs with EBOV and SUDV infections, and proved to be protective [58]. When supplemented with MR191, an anti-MARV mAb, the triple mAb cocktail exhibited full protection against death in MARV-infected NHPs [58]. In addition, antibody cocktail RIID F6-H2 is comprised of two SUDV specific mAbs, 16F6 and X10H2, targeting the base and glycan cap of GP, respectively [100,101]. This mAb cocktail protects macaques from the SUDV challenge with two doses on day 4 and day 6, at 25 mg/kg per mAb. The model is not fully lethal; 50% of the mock-treated exposed control animals survived the SUDV challenge [100].

A second cocktail named MBP134^{AF} contains two non-competing IFL targeting mAbs, ADI-15878 (described in IFL region mAb section) and ADI-23774, which was selected after specificity maturation of ADI-15946 to bind SUDV GP using yeast-display technology [102,103]. The mAb pair was further optimized to improve their capacity to activate NK cell functions by adopting all afucosylated glycans (thus the AF in the cocktail name), in order to reach higher efficacy against EBOV [31,102]. The cocktail protects NHPs against EBOV, SUDV, and BDBV [59].

A third cocktail, which also comprises two mAbs, rEBOV-442 and rEBOV-515, was recently reported to protect NHP from disease caused by EBOV, BDBV, and SUDV [61]. These two mAbs exhibited synergy in neutralization by occupying non-overlapping epitopes, with rEBOV-442 targeting the glycan cap region [75], and rEBOV-515 targeting the conserved IFL region. Compared to the previously described ADI-15946 [89] and EBOV-520 [60], the footprint of rEBOV-515 is more conserved and thus provides better neutralizing breath against SUDV [61].

The fourth cocktail of two human survivor antibodies was recently described [62]. This cocktail includes antibodies isolated from survivors of EVD: 1C3 and 1C11, and also protects NHP against lethal challenge with EBOV or SUDV [62]. The 1C3 and 1C11 pair was chosen from a broad analysis of the VIC consortium, and has been tested in multiple animal models (mouse, guinea pig, and NHP). 1C3 uniquely targets the head region with one Fab anchoring into the GP chalice to bind all the three monomers of the GP trimer simultaneously (Figure 3F). This tripartite recognition mode leads to strong binding to the GP trimer, and no cross-reactivity to the dimeric shed sGP. The GP specificity of 1C3 is unique for an EBOV GP head-binding antibody and results from its particular quaternary epitope recognition. Interestingly, different parts of 1C3 target the identical GP residues on each monomer. For example, GP residues D117 in monomer A forms hydrogen bonds with CDRH3 of 1C3, in monomer B forms hydrogen bonds to FRL3, and contacts FRL1 in monomer C [62]. The broadly reactive 1C11 antibody targets the fusion loop/base region via an epitope similar to that of 6D6 [88] and ADI-15878 [85]. 1C11 binds with three copies of the Fab per GP trimer, with each individual Fab bridging two adjacent monomers together to link the fusion loop paddle of monomer A to the neighboring monomer B, including the N-linked glycan at position 563 at each of the three positions around the trimer.

These third-generation candidate therapeutic cocktails can all protect NHP against infection by multiple ebolaviruses, representing the direction of therapeutic development in the field.

For the two approved EVD treatments, the mAb114 monotherapy required a 50 mg/kg dose, whereas the REGN-EB3 triple cocktail required 50 mg/kg of each mAb component for a total 150 mg/kg dose [21,22]. Here we also compare the recently reported cocktails that are protective in NHPs. FVM04/CA45 was protective against EBOV when offered at 40 mg/kg (20 mg/kg each) on day 4, and protective against SUDV when offered at 40 mg/kg (20 mg/kg each) on day 4, plus a second dose at 13 mg/kg (8 mg/kg FVM04, 5 mg/kg CA45) on day 6. Against MARV, MR191 was administered in addition to the two-mAb FVM04/CA45 cocktail at 50 mg/kg, with the first dose (90 mg/kg total) on day 4 and the second dose on day 6 (70 mg/kg total). The second cocktail, MBP134^{AF}, was tested in NHPs against EBOV, SUDV, and BDBV, and showed protection with a single 25 mg/kg dose. However, in both studies, the mock-treated exposed control animals survived the SUDV challenge (one out of two in the FVM04/CA45 study, two out of four in the MBP134^{AF} study), which limits the significance of the protection results. In the MBP134^{AF} BDBV protection study, five out of six treated animals survived. The third cocktail, rEBOV-442/515, was tested in NHPs against EBOV, SUDV, and BDBV, and showed protection with a two-dose regimen at 30 mg/kg (10 mg/kg rEBOV-442, 20 mg/kg rEBOV-515). The fourth cocktail, 1C3/1C11, was protective against EBOV at 25 mg/kg, and protective against SUDV at 50 mg/kg, both with two doses on day 4 and day 7. The synergistic effect between mAbs described in several studies also hinted at the possibility that these dosages could be further optimized.

One limitation shared by these NHP studies is the relatively small number of animals per group, which results in lower statistical power for examination of the significance of the beneficial effect. Future studies will need to include additional NHPs to test the reported regimens and dosages and to explore the efficacy of single and lower dosages of the proposed cocktails.

4. Conclusions

After the neutralizing monoclonal antibody KZ52 was found not to protect NHPs infected with Ebola, it was initially thought to be an indication that mAbs may not be effective against rapidly progressing EVD [104]. The discovery that the three non-neutralizing antibodies of MB-003 could protect primates, and subsequent refinement and improvement of antibody cocktails to include neutralizing antibodies that targeted epitopes in the GP base allowed not only survival, but reversion of advanced disease symptoms, as demonstrated by ZMapp in NHPs [13], set a starting point for use of mAbs as therapeutics. The broad collaborative analysis of the VIC illuminated multiple antibody features that led to protection and proposed that antibody therapies ideally should offer potent neutralization, a lack of an un-neutralized viral fraction, and recruitment of Fc effector functions. Further, the analysis by the VIC indicated that the Fc effector function recruitment, particularly phagocytosis, could be strongest at the top of the molecule (e.g., head, glycan cap, and MLD), and that the head epitope, in particular, was a sweet spot that permits both effector function recruitment, as well as potent neutralization by blocking receptor binding [31]. The VIC work established standards, enabled cross-comparison, and allowed side-by-side evaluation of all the cocktail components thus far, including the second-generation mAb cocktails (MB-003, ZMAb, ZMapp, MBP134^{AF}, REGN-EB3), and the proposed mAb cocktail (mAb 1C3 and 1C11) for the combination of complementary activities with antibodies of broad specificity.

Second-generation antibody treatments combine both neutralization and Fc functions, either by binding a single monotherapy at the head sweet spot (i.e., mAb114 [14]), or by combining antibodies having different epitopes and functions, as in REGN-EB3 [20]. The third generation of antibody treatments aims to confer protection against other disease-causing ebolaviruses, and ideally, involve a lower dosage or a simpler therapeutic regimen. These cocktails, including MBP134^{AF}, rEBOV-520/548, rEBOV-442/515, and 1C3/1C11, have reduced the three-antibody cocktail to two, and each contains at least one mAb targeting the fusion loop region, with the other mAb binding to IFL, glycan cap, or the apex/head region [59–62]. Mapping the structures and activities of antibodies effective against different ebolaviruses illuminates not only emergency post-exposure treatment options, but also illustrates the types of antibodies that vaccines should elicit.

The role of sGP in the immune system during viral infection still remains largely unclear. In the VIC systematic review, sGP cross-reactivity did not significantly affect in vivo protection [31]. The approved monotherapy mAb114 cross-reacts with sGP, and protects against EVD in NHP studies and in the clinical trial at a 50 mg/kg dosage. However, the currently lowest treatment dosage to protect NHP, 25 mg/kg, was achieved by mAb cocktails MBP134^{AF} and 1C3/1C11, the components of which are specific to the GP trimer and do not bind to sGP [59,62]. This result could suggest that a lower effective dosage could be achieved using GP-specific antibodies alone. More studies are needed to better understand whether sGP is an immunogen or a decoy, so that we can address whether cross-reactivity against sGP is an advantageous feature or should be avoided for mAb candidates.

With the current vaccine only available for EBOV, there remains the need for the continued availability of antibodies as it is impractical to vaccinate all, and the frequency and unpredictable timing and location of outbreaks suggest continued treatment development is needed. Studies focusing on characterizing vaccine-elicited mAbs and comparison to those elicited from viral infection would also contribute to the next stage of broadly effective vaccine and immunotherapy development [56,57].

Lastly, platforms and expertise honed on the Ebola virus and the collaborative framework of the VIC [31] were both deployed in rapid development, advancement, and comparison of antibody therapeutics against SARS-CoV-2 [105], and will likely be called upon again against future emerging infections.

Author Contributions: Writing—original draft preparation, X.Y., E.O.S.; writing—review and editing, X.Y., E.O.S.; visualization, X.Y.; All authors have read and agreed to the published version of the manuscript.

Funding: This research was funded by the National Institute of Allergy and Infectious Diseases (NIAID) U19 AI142790 and the National Institute of Health (NIH) R01 AI132204.

Institutional Review Board Statement: Not applicable.

Informed Consent Statement: Not applicable.

Data Availability Statement: Not applicable.

Acknowledgments: We gratefully acknowledge our funding from the National Institute of Allergy and Infectious Diseases (NIAID) U19 AI142790 and from the National Institute of Health (NIH) R01 AI132204. Figures 1 and 2B,D were created with BioRender.com.

Conflicts of Interest: The authors declare no conflict of interest.

References

1. Yang, X.-L.; Tan, C.W.; Anderson, D.E.; Jiang, R.-D.; Li, B.; Zhang, W.; Zhu, Y.; Lim, X.F.; Zhou, P.; Liu, X.-L.; et al. Characterization of a Filovirus (Měnglà Virus) from Rousettus Bats in China. *Nat. Microbiol.* **2019**, *4*, 390–395. [CrossRef] [PubMed]
2. Shi, M.; Lin, X.-D.; Chen, X.; Tian, J.-H.; Chen, L.-J.; Li, K.; Wang, W.; Eden, J.-S.; Shen, J.-J.; Liu, L.; et al. The Evolutionary History of Vertebrate RNA Viruses. *Nature* **2018**, *556*, 197–202. [CrossRef] [PubMed]
3. Brauburger, K.; Hume, A.J.; Mühlberger, E.; Olejnik, J. Forty-Five Years of Marburg Virus Research. *Viruses* **2012**, *4*, 1878–1927. [CrossRef] [PubMed]
4. World Health Organization. Ebola Haemorrhagic Fever in Zaire, 1976. *Bull. World Health Organ.* **1978**, *56*, 271–293.
5. World Health Organization. Ebola Haemorrhagic Fever in Sudan, 1976. Report of a WHO/International Study Team. *Bull. World Health Organ.* **1978**, *56*, 247–270.
6. CDC. 2021 Democratic Republic of the Congo, North Kivu Province. Available online: https://www.cdc.gov/vhf/ebola/outbreaks/drc/2021-february.html (accessed on 3 May 2021).
7. 2021 Guinea, N'Zérékoré Prefecture. Available online: https://www.cdc.gov/vhf/ebola/outbreaks/guinea/2021-february.html (accessed on 3 May 2021).
8. CDC Brief Report: Outbreak of Marburg Virus Hemorrhagic Fever—Angola, 1 October 2004–29 March 2005. Available online: https://www.cdc.gov/mmwr/preview/mmwrhtml/mm54d330a1.htm (accessed on 18 May 2021).
9. History of Ebola Virus Disease (EVD) Outbreaks. Available online: https://www.cdc.gov/vhf/ebola/history/chronology.html (accessed on 18 May 2021).
10. Mehedi, M.; Groseth, A.; Feldmann, H.; Ebihara, H. Clinical Aspects of Marburg Hemorrhagic Fever. *Future Virol.* **2011**, *6*, 1091–1106. [CrossRef] [PubMed]
11. Goeijenbier, M.; van Kampen, J.J.A.; Reusken, C.B.E.M.; Koopmans, M.P.G.; van Gorp, E.C.M. Ebola Virus Disease: A Review on Epidemiology, Symptoms, Treatment and Pathogenesis. *Neth. J. Med.* **2014**, *72*, 442–448. [PubMed]
12. Litterman, N.; Lipinski, C.; Ekins, S. Small Molecules with Antiviral Activity against the Ebola Virus. *F1000Research* **2015**, *4*, 38. [CrossRef] [PubMed]
13. Qiu, X.; Wong, G.; Audet, J.; Bello, A.; Fernando, L.; Alimonti, J.B.; Fausther-Bovendo, H.; Wei, H.; Aviles, J.; Hiatt, E.; et al. Reversion of Advanced Ebola Virus Disease in Nonhuman Primates with ZMapp. *Nature* **2014**, *514*, 47–53. [CrossRef]
14. Corti, D.; Misasi, J.; Mulangu, S.; Stanley, D.A.; Kanekiyo, M.; Wollen, S.; Ploquin, A.; Doria-Rose, N.A.; Staupe, R.P.; Bailey, M.; et al. Protective Monotherapy against Lethal Ebola Virus Infection by a Potently Neutralizing Antibody. *Science* **2016**, *351*, 1339–1342. [CrossRef]

15. Henao-Restrepo, A.M.; Camacho, A.; Longini, I.M.; Watson, C.H.; Edmunds, W.J.; Egger, M.; Carroll, M.W.; Dean, N.E.; Diatta, I.; Doumbia, M.; et al. Efficacy and Effectiveness of an RVSV-Vectored Vaccine in Preventing Ebola Virus Disease: Final Results from the Guinea Ring Vaccination, Open-Label, Cluster-Randomised Trial (Ebola Ça Suffit!). *Lancet* **2017**, *389*, 505–518. [CrossRef]
16. Bache, B.E.; Grobusch, M.P.; Agnandji, S.T. Safety, Immunogenicity and Risk-Benefit Analysis of RVSV-ΔG-ZEBOV-GP (V920) Ebola Vaccine in Phase I-III Clinical Trials across Regions. *Future Microbiol.* **2020**, *15*, 85–106. [CrossRef] [PubMed]
17. Ollmann Saphire, E. A Vaccine against Ebola Virus. *Cell* **2020**, *181*, 6. [CrossRef] [PubMed]
18. Czarska-Thorley, D. New Vaccine for Prevention of Ebola Virus Disease Recommended for Approval in the European Union. Available online: https://www.ema.europa.eu/en/news/new-vaccine-prevention-ebola-virus-disease-recommended-approval-european-union (accessed on 13 July 2021).
19. Pollard, A.J.; Launay, O.; Lelievre, J.-D.; Lacabaratz, C.; Grande, S.; Goldstein, N.; Robinson, C.; Gaddah, A.; Bockstal, V.; Wiedemann, A.; et al. Safety and Immunogenicity of a Two-Dose Heterologous Ad26. ZEBOV and MVA-BN-Filo Ebola Vaccine Regimen in Adults in Europe (EBOVAC2): A Randomised, Observer-Blind, Participant-Blind, Placebo-Controlled, Phase 2 Trial. *Lancet Infect. Dis.* **2021**, *21*, 493–506. [CrossRef]
20. Mulangu, S.; Dodd, L.E.; Davey, R.T., Jr.; Tshiani Mbaya, O.; Proschan, M.; Mukadi, D.; Lusakibanza Manzo, M.; Nzolo, D.; Tshomba Oloma, A.; Ibanda, A.; et al. A Randomized, Controlled Trial of Ebola Virus Disease Therapeutics. *N. Engl. J. Med.* **2019**, *381*, 2293–2303. [CrossRef] [PubMed]
21. Food and Drug Administration. FDA Approves First Treatment for Ebola Virus. Available online: https://www.fda.gov/news-events/press-announcements/fda-approves-first-treatment-ebola-virus (accessed on 19 May 2021).
22. Food and Drug Administration. FDA Approves Treatment for Ebola Virus. Available online: https://www.fda.gov/drugs/drug-safety-and-availability/fda-approves-treatment-ebola-virus (accessed on 19 May 2021).
23. Sanchez, A.; Trappier, S.G.; Mahy, B.W.; Peters, C.J.; Nichol, S.T. The Virion Glycoproteins of Ebola Viruses Are Encoded in Two Reading Frames and Are Expressed through Transcriptional Editing. *Proc. Natl. Acad. Sci. USA* **1996**, *93*, 3602–3607. [CrossRef] [PubMed]
24. Sanchez, A.; Yang, Z.Y.; Xu, L.; Nabel, G.J.; Crews, T.; Peters, C.J. Biochemical Analysis of the Secreted and Virion Glycoproteins of Ebola Virus. *J. Virol.* **1998**, *72*, 6442–6447. [CrossRef] [PubMed]
25. Lee, J.E.; Fusco, M.L.; Hessell, A.J.; Oswald, W.B.; Burton, D.R.; Saphire, E.O. Structure of the Ebola Virus Glycoprotein Bound to an Antibody from a Human Survivor. *Nature* **2008**, *454*, 177–182. [CrossRef]
26. Lee, J.E.; Saphire, E.O. Ebolavirus Glycoprotein Structure and Mechanism of Entry. *Future Virol.* **2009**, *4*, 621–635. [CrossRef]
27. Volchkov, V.E.; Becker, S.; Volchkova, V.A.; Ternovoj, V.A.; Kotov, A.N.; Netesov, S.V.; Klenk, H.-D. GP MRNA of Ebola Virus Is Edited by the Ebola Virus Polymerase and by T7 and Vaccinia Virus Polymerases1. *Virology* **1995**, *214*, 421–430. [CrossRef]
28. De La Vega, M.-A.; Wong, G.; Kobinger, G.P.; Qiu, X. The Multiple Roles of SGP in Ebola Pathogenesis. *Viral Immunol.* **2015**, *28*, 3–9. [CrossRef] [PubMed]
29. Pallesen, J.; Murin, C.D.; de Val, N.; Cottrell, C.A.; Hastie, K.M.; Turner, H.L.; Fusco, M.L.; Flyak, A.I.; Zeitlin, L.; Crowe, J.E.; et al. Structures of Ebola Virus GP and SGP in Complex with Therapeutic Antibodies. *Nat. Microbiol.* **2016**, *1*, 1–9. [CrossRef] [PubMed]
30. Davis, C.W.; Jackson, K.J.L.; McElroy, A.K.; Halfmann, P.; Huang, J.; Chennareddy, C.; Piper, A.E.; Leung, Y.; Albariño, C.G.; Crozier, I.; et al. Longitudinal Analysis of the Human B Cell Response to Ebola Virus Infection. *Cell* **2019**, *177*, 1566–1582.e17. [CrossRef] [PubMed]
31. Saphire, E.O.; Schendel, S.L.; Fusco, M.L.; Gangavarapu, K.; Gunn, B.M.; Wec, A.Z.; Halfmann, P.J.; Brannan, J.M.; Herbert, A.S.; Qiu, X.; et al. Systematic Analysis of Monoclonal Antibodies against Ebola Virus GP Defines Features That Contribute to Protection. *Cell* **2018**, *174*, 938–952.e13. [CrossRef]
32. Sanchez, A.; Geisbert, T.W.; Feldmann, H. Filoviridae: Marburg and Ebola Viruses. In *Fields Virology*; Lippincot Williams & Wilkins: Philadelphia, PA, USA, 2007.
33. Jeffers, S.A.; Sanders, D.A.; Sanchez, A. Covalent Modifications of the Ebola Virus Glycoprotein. *J. Virol.* **2002**, *76*, 12463–12472. [CrossRef] [PubMed]
34. Volchkov, V.E.; Volchkova, V.A.; Ströher, U.; Becker, S.; Dolnik, O.; Cieplik, M.; Garten, W.; Klenk, H.D.; Feldmann, H. Proteolytic Processing of Marburg Virus Glycoprotein. *Virology* **2000**, *268*, 1–6. [CrossRef] [PubMed]
35. Fusco, M.L.; Hashiguchi, T.; Cassan, R.; Biggins, J.E.; Murin, C.D.; Warfield, K.L.; Li, S.; Holtsberg, F.W.; Shulenin, S.; Vu, H.; et al. Protective MAbs and Cross-Reactive MAbs Raised by Immunization with Engineered Marburg Virus GPs. *PLoS Pathogens* **2015**, *11*, e1005016.
36. Nanbo, A.; Imai, M.; Watanabe, S.; Noda, T.; Takahashi, K.; Neumann, G.; Halfmann, P.; Kawaoka, Y. Ebolavirus Is Internalized into Host Cells via Macropinocytosis in a Viral Glycoprotein-Dependent Manner. *PLoS Pathog.* **2010**, *6*, e1001121. [CrossRef] [PubMed]
37. Saeed, M.F.; Kolokoltsov, A.A.; Albrecht, T.; Davey, R.A. Cellular Entry of Ebola Virus Involves Uptake by a Macropinocytosis-like Mechanism and Subsequent Trafficking through Early and Late Endosomes. *PLoS Pathog.* **2010**, *6*, e1001110. [CrossRef] [PubMed]
38. Aleksandrowicz, P.; Marzi, A.; Biedenkopf, N.; Beimforde, N.; Becker, S.; Hoenen, T.; Feldmann, H.; Schnittler, H.-J. Ebola Virus Enters Host Cells by Macropinocytosis and Clathrin-Mediated Endocytosis. *J. Infect. Dis.* **2011**, *204* (Suppl. 3), S957–S967. [CrossRef]
39. Chandran, K.; Sullivan, N.J.; Felbor, U.; Whelan, S.P.; Cunningham, J.M. Endosomal Proteolysis of the Ebola Virus Glycoprotein Is Necessary for Infection. *Science* **2005**, *308*, 1643–1645. [CrossRef] [PubMed]

40. Schornberg, K.; Matsuyama, S.; Kabsch, K.; Delos, S.; Bouton, A.; White, J. Role of Endosomal Cathepsins in Entry Mediated by the Ebola Virus Glycoprotein. *J. Virol.* **2006**, *80*, 4174–4178. [CrossRef] [PubMed]

41. Kaletsky, R.L.; Simmons, G.; Bates, P. Proteolysis of the Ebola Virus Glycoproteins Enhances Virus Binding and Infectivity. *J. Virol.* **2007**, *81*, 13378–13384. [CrossRef] [PubMed]

42. Wang, H.; Shi, Y.; Song, J.; Qi, J.; Lu, G.; Yan, J.; Gao, G.F. Ebola Viral Glycoprotein Bound to Its Endosomal Receptor Niemann-Pick C1. *Cell* **2016**, *164*, 258–268. [CrossRef] [PubMed]

43. Dube, D.; Brecher, M.B.; Delos, S.E.; Rose, S.C.; Park, E.W.; Schornberg, K.L.; Kuhn, J.H.; White, J.M. The Primed Ebolavirus Glycoprotein (19-Kilodalton GP1,2): Sequence and Residues Critical for Host Cell Binding. *J. Virol.* **2009**, *83*, 2883–2891. [CrossRef] [PubMed]

44. Huse, W.D.; Sastry, L.; Iverson, S.A.; Kang, A.S.; Alting-Mees, M.; Burton, D.R.; Benkovic, S.J.; Lerner, R.A. Generation of a Large Combinatorial Library of the Immunoglobulin Repertoire in Phage Lambda. *Science* **1989**, *246*, 1275–1281. [CrossRef]

45. McCafferty, J.; Griffiths, A.D.; Winter, G.; Chiswell, D.J. Phage Antibodies: Filamentous Phage Displaying Antibody Variable Domains. *Nature* **1990**, *348*, 552–554. [CrossRef] [PubMed]

46. Maruyama, T.; Rodriguez, L.L.; Jahrling, P.B.; Sanchez, A.; Khan, A.S.; Nichol, S.T.; Peters, C.J.; Parren, P.W.; Burton, D.R. Ebola Virus Can Be Effectively Neutralized by Antibody Produced in Natural Human Infection. *J. Virol.* **1999**, *73*, 6024–6030. [CrossRef] [PubMed]

47. Wilson, J.A.; Hevey, M.; Bakken, R.; Guest, S.; Bray, M.; Schmaljohn, A.L.; Hart, M.K. Epitopes Involved in Antibody-Mediated Protection from Ebola Virus. *Science* **2000**, *287*, 1664–1666. [CrossRef] [PubMed]

48. Qiu, X.; Audet, J.; Wong, G.; Pillet, S.; Bello, A.; Cabral, T.; Strong, J.E.; Plummer, F.; Corbett, C.R.; Alimonti, J.B.; et al. Successful Treatment of Ebola Virus-Infected Cynomolgus Macaques with Monoclonal Antibodies. *Sci. Transl. Med.* **2012**, *4*, ra81–ra138. [CrossRef]

49. Olinger, G.G., Jr.; Pettitt, J.; Kim, D.; Working, C.; Bohorov, O.; Bratcher, B.; Hiatt, E.; Hume, S.D.; Johnson, A.K.; Morton, J.; et al. Delayed Treatment of Ebola Virus Infection with Plant-Derived Monoclonal Antibodies Provides Protection in Rhesus Macaques. *Proc. Natl. Acad. Sci. USA* **2012**, *109*, 18030–18035. [CrossRef] [PubMed]

50. Keck, Z.-Y.; Enterlein, S.G.; Howell, K.A.; Vu, H.; Shulenin, S.; Warfield, K.L.; Froude, J.W.; Araghi, N.; Douglas, R.; Biggins, J.; et al. Macaque Monoclonal Antibodies Targeting Novel Conserved Epitopes within Filovirus Glycoprotein. *J. Virol.* **2016**, *90*, 279–291. [CrossRef] [PubMed]

51. Furuyama, W.; Marzi, A.; Nanbo, A.; Haddock, E.; Maruyama, J.; Miyamoto, H.; Igarashi, M.; Yoshida, R.; Noyori, O.; Feldmann, H.; et al. Discovery of an Antibody for Pan-Ebolavirus Therapy. *Sci. Rep.* **2016**, *6*, 20514. [CrossRef] [PubMed]

52. Pascal, K.E.; Dudgeon, D.; Trefry, J.C.; Anantpadma, M.; Sakurai, Y.; Murin, C.D.; Turner, H.L.; Fairhurst, J.; Torres, M.; Rafique, A.; et al. Development of Clinical-Stage Human Monoclonal Antibodies That Treat Advanced Ebola Virus Disease in Nonhuman Primates. *J. Infect. Dis.* **2018**, *218*, S612–S626. [CrossRef] [PubMed]

53. Bornholdt, Z.A.; Turner, H.L.; Murin, C.D.; Li, W.; Sok, D.; Souders, C.A.; Piper, A.E.; Goff, A.; Shamblin, J.D.; Wollen, S.E.; et al. Isolation of Potent Neutralizing Antibodies from a Survivor of the 2014 Ebola Virus Outbreak. *Science* **2016**, *351*, 1078–1083. [CrossRef] [PubMed]

54. Flyak, A.I.; Shen, X.; Murin, C.D.; Turner, H.L.; David, J.A.; Fusco, M.L.; Lampley, R.; Kose, N.; Ilinykh, P.A.; Kuzmina, N.; et al. Cross-Reactive and Potent Neutralizing Antibody Responses in Human Survivors of Natural Ebolavirus Infection. *Cell* **2016**, *164*, 392–405. [CrossRef] [PubMed]

55. Wec, A.Z.; Herbert, A.S.; Murin, C.D.; Nyakatura, E.K.; Abelson, D.M.; Fels, J.M.; He, S.; James, R.M.; de La Vega, M.-A.; Zhu, W.; et al. Antibodies from a Human Survivor Define Sites of Vulnerability for Broad Protection against Ebolaviruses. *Cell* **2017**, *169*, 878–890.e15. [CrossRef]

56. Ehrhardt, S.A.; Zehner, M.; Krähling, V.; Cohen-Dvashi, H.; Kreer, C.; Elad, N.; Gruell, H.; Ercanoglu, M.S.; Schommers, P.; Gieselmann, L.; et al. Polyclonal and Convergent Antibody Response to Ebola Virus Vaccine RVSV-ZEBOV. *Nat. Med.* **2019**, *25*, 1589–1600. [CrossRef] [PubMed]

57. Rijal, P.; Elias, S.C.; Machado, S.R.; Xiao, J.; Schimanski, L.; O'Dowd, V.; Baker, T.; Barry, E.; Mendelsohn, S.C.; Cherry, C.J.; et al. Therapeutic Monoclonal Antibodies for Ebola Virus Infection Derived from Vaccinated Humans. *Cell Rep.* **2019**, *27*, 172–186.e7. [CrossRef]

58. Brannan, J.M.; He, S.; Howell, K.A.; Prugar, L.I.; Zhu, W.; Vu, H.; Shulenin, S.; Kailasan, S.; Raina, H.; Wong, G.; et al. Post-Exposure Immunotherapy for Two Ebolaviruses and Marburg Virus in Nonhuman Primates. *Nat. Commun.* **2019**, *10*, 105. [CrossRef]

59. Bornholdt, Z.A.; Herbert, A.S.; Mire, C.E.; He, S.; Cross, R.W.; Wec, A.Z.; Abelson, D.M.; Geisbert, J.B.; James, R.M.; Rahim, M.N.; et al. A Two-Antibody Pan-Ebolavirus Cocktail Confers Broad Therapeutic Protection in Ferrets and Nonhuman Primates. *Cell Host Microbe* **2019**, *25*, 49–58.e5. [CrossRef] [PubMed]

60. Gilchuk, P.; Murin, C.D.; Milligan, J.C.; Cross, R.W.; Mire, C.E.; Ilinykh, P.A.; Huang, K.; Kuzmina, N.; Altman, P.X.; Hui, S.; et al. Analysis of a Therapeutic Antibody Cocktail Reveals Determinants for Cooperative and Broad Ebolavirus Neutralization. *Immunity* **2020**, *52*, 388–403.e12. [CrossRef]

61. Gilchuk, P.; Murin, C.D.; Cross, R.W.; Ilinykh, P.A.; Huang, K.; Kuzmina, N.; Borisevich, V.; Agans, K.N.; Geisbert, J.B.; Zost, S.J.; et al. Pan-Ebolavirus Protective Therapy by Two Multifunctional Human Antibodies. *Cell* **2021**, *184*, 5593–5607.e18. [CrossRef]

62. Milligan, J.C.; Davis, C.W.; Yu, X.; Ilinykh, P.A.; Huang, K.; Halfmann, P.J.; Cross, R.W.; Borisevich, V.; Agans, K.N.; Geisbert, J.B.; et al. Asymmetric and Non-Stoichiometric Glycoprotein Recognition by Two Distinct Antibodies Results in Broad Protection against Ebolaviruses. *Cell* **2022**, *185*, 995–1007.e18. [CrossRef]

63. Honnold, S.P.; Bakken, R.R.; Fisher, D.; Lind, C.M.; Cohen, J.W.; Eccleston, L.T.; Spurgers, K.B.; Maheshwari, R.K.; Glass, P.J. Second Generation Inactivated Eastern Equine Encephalitis Virus Vaccine Candidates Protect Mice against a Lethal Aerosol Challenge. *PLoS ONE* **2014**, *9*, e104708. [CrossRef] [PubMed]

64. Halfmann, P.; Kim, J.H.; Ebihara, H.; Noda, T.; Neumann, G.; Feldmann, H.; Kawaoka, Y. Generation of Biologically Contained Ebola Viruses. *Proc. Natl. Acad. Sci. USA* **2008**, *105*, 1129–1133. [CrossRef] [PubMed]

65. Wong, A.C.; Sandesara, R.G.; Mulherkar, N.; Whelan, S.P.; Chandran, K. A Forward Genetic Strategy Reveals Destabilizing Mutations in the Ebolavirus Glycoprotein That Alter Its Protease Dependence during Cell Entry. *J. Virol.* **2010**, *84*, 163–175. [CrossRef] [PubMed]

66. Ilinykh, P.A.; Shen, X.; Flyak, A.I.; Kuzmina, N.; Ksiazek, T.G.; Crowe, J.E., Jr.; Bukreyev, A. Chimeric Filoviruses for Identification and Characterization of Monoclonal Antibodies. *J. Virol.* **2016**, *90*, 3890–3901. [CrossRef] [PubMed]

67. Wec, A.Z.; Nyakatura, E.K.; Herbert, A.S.; Howell, K.A.; Holtsberg, F.W.; Bakken, R.R.; Mittler, E.; Christin, J.R.; Shulenin, S.; Jangra, R.K.; et al. A "Trojan Horse" Bispecific-Antibody Strategy for Broad Protection against Ebolaviruses. *Science* **2016**, *354*, 350–354. [CrossRef] [PubMed]

68. Ng, M.; Ndungo, E.; Jangra, R.K.; Cai, Y.; Postnikova, E.; Radoshitzky, S.R.; Dye, J.M.; Ramírez de Arellano, E.; Negredo, A.; Palacios, G.; et al. Cell Entry by a Novel European Filovirus Requires Host Endosomal Cysteine Proteases and Niemann-Pick C1. *Virology* **2014**, *468–470*, 637–646. [CrossRef] [PubMed]

69. Gunn, B.M.; Yu, W.-H.; Karim, M.M.; Brannan, J.M.; Herbert, A.S.; Wec, A.Z.; Halfmann, P.J.; Fusco, M.L.; Schendel, S.L.; Gangavarapu, K.; et al. A Role for Fc Function in Therapeutic Monoclonal Antibody-Mediated Protection against Ebola Virus. *Cell Host Microbe* **2018**, *24*, 221–233.e5. [CrossRef] [PubMed]

70. Bournazos, S.; DiLillo, D.J.; Goff, A.J.; Glass, P.J.; Ravetch, J.V. Differential Requirements for FcγR Engagement by Protective Antibodies against Ebola Virus. *Proc. Natl. Acad. Sci. USA* **2019**, *116*, 20054–20062. [CrossRef] [PubMed]

71. Gunn, B.M.; Lu, R.; Slein, M.D.; Ilinykh, P.A.; Huang, K.; Atyeo, C.; Schendel, S.L.; Kim, J.; Cain, C.; Roy, V.; et al. A Fc Engineering Approach to Define Functional Humoral Correlates of Immunity against Ebola Virus. *Immunity* **2021**, *54*, 815–828.e5. [CrossRef]

72. Murin, C.D.; Fusco, M.L.; Bornholdt, Z.A.; Qiu, X.; Olinger, G.G.; Zeitlin, L.; Kobinger, G.P.; Ward, A.B.; Saphire, E.O. Structures of Protective Antibodies Reveal Sites of Vulnerability on Ebola Virus. *Proc. Natl. Acad. Sci. USA* **2014**, *111*, 17182–17187. [CrossRef] [PubMed]

73. Flyak, A.I.; Kuzmina, N.; Murin, C.D.; Bryan, C.; Davidson, E.; Gilchuk, P.; Gulka, C.P.; Ilinykh, P.A.; Shen, X.; Huang, K.; et al. Broadly Neutralizing Antibodies from Human Survivors Target a Conserved Site in the Ebola Virus Glycoprotein HR2-MPER Region. *Nat. Microbiol.* **2018**, *3*, 670–677. [CrossRef] [PubMed]

74. Davidson, E.; Bryan, C.; Fong, R.H.; Barnes, T.; Pfaff, J.M.; Mabila, M.; Rucker, J.B.; Doranz, B.J. Mechanism of Binding to Ebola Virus Glycoprotein by the ZMapp, ZMAb, and MB-003 Cocktail Antibodies. *J. Virol.* **2015**, *89*, 10982–10992. [CrossRef] [PubMed]

75. Murin, C.D.; Gilchuk, P.; Ilinykh, P.A.; Huang, K.; Kuzmina, N.; Shen, X.; Bruhn, J.F.; Bryan, A.L.; Davidson, E.; Doranz, B.J.; et al. Convergence of a Common Solution for Broad Ebolavirus Neutralization by Glycan Cap-Directed Human Antibodies. *Cell Rep.* **2021**, *35*, 108984. [CrossRef] [PubMed]

76. Zhang, Q.; Gui, M.; Niu, X.; He, S.; Wang, R.; Feng, Y.; Kroeker, A.; Zuo, Y.; Wang, H.; Wang, Y.; et al. Potent Neutralizing Monoclonal Antibodies against Ebola Virus Infection. *Sci. Rep.* **2016**, *6*, 25856. [CrossRef] [PubMed]

77. Flyak, A.I.; Ilinykh, P.A.; Murin, C.D.; Garron, T.; Shen, X.; Fusco, M.L.; Hashiguchi, T.; Bornholdt, Z.A.; Slaughter, J.C.; Sapparapu, G.; et al. Mechanism of Human Antibody-Mediated Neutralization of Marburg Virus. *Cell* **2015**, *160*, 893–903. [CrossRef] [PubMed]

78. Hashiguchi, T.; Fusco, M.L.; Bornholdt, Z.A.; Lee, J.E.; Flyak, A.I.; Matsuoka, R.; Kohda, D.; Yanagi, Y.; Hammel, M.; Crowe, J.E., Jr.; et al. Structural Basis for Marburg Virus Neutralization by a Cross-Reactive Human Antibody. *Cell* **2015**, *160*, 904–912. [CrossRef]

79. King, L.B.; Fusco, M.L.; Flyak, A.I.; Ilinykh, P.A.; Huang, K.; Gunn, B.; Kirchdoerfer, R.N.; Hastie, K.M.; Sangha, A.K.; Meiler, J.; et al. The Marburgvirus-Neutralizing Human Monoclonal Antibody MR191 Targets a Conserved Site to Block Virus Receptor Binding. *Cell Host Microbe* **2018**, *23*, 101–109.e4. [CrossRef] [PubMed]

80. Misasi, J.; Gilman, M.S.A.; Kanekiyo, M.; Gui, M.; Cagigi, A.; Mulangu, S.; Corti, D.; Ledgerwood, J.E.; Lanzavecchia, A.; Cunningham, J.; et al. Structural and Molecular Basis for Ebola Virus Neutralization by Protective Human Antibodies. *Science* **2016**, *351*, 1343–1346. [CrossRef] [PubMed]

81. Howell, K.A.; Qiu, X.; Brannan, J.M.; Bryan, C.; Davidson, E.; Holtsberg, F.W.; Wec, A.Z.; Shulenin, S.; Biggins, J.E.; Douglas, R.; et al. Antibody Treatment of Ebola and Sudan Virus Infection via a Uniquely Exposed Epitope within the Glycoprotein Receptor-Binding Site. *Cell Rep.* **2016**, *15*, 1514–1526. [CrossRef]

82. Cohen-Dvashi, H.; Zehner, M.; Ehrhardt, S.; Katz, M.; Elad, N.; Klein, F.; Diskin, R. Structural Basis for a Convergent Immune Response against Ebola Virus. *Cell Host Microbe* **2020**, *27*, 418–427.e4. [CrossRef]

83. Gorman, J.; Chuang, G.-Y.; Lai, Y.-T.; Shen, C.-H.; Boyington, J.C.; Druz, A.; Geng, H.; Louder, M.K.; McKee, K.; Rawi, R.; et al. Structure of Super-Potent Antibody CAP256-VRC26.25 in Complex with HIV-1 Envelope Reveals a Combined Mode of Trimer-Apex Recognition. *Cell Rep.* **2020**, *31*, 107488. [CrossRef] [PubMed]

84. King, L.B.; Milligan, J.C.; West, B.R.; Schendel, S.L.; Ollmann Saphire, E. Achieving Cross-Reactivity with Pan-Ebolavirus Antibodies. *Curr. Opin. Virol.* **2019**, *34*, 140–148. [CrossRef] [PubMed]

85. West, B.R.; Moyer, C.L.; King, L.B.; Fusco, M.L.; Milligan, J.C.; Hui, S.; Saphire, E.O. Structural Basis of Pan-Ebolavirus Neutralization by a Human Antibody against a Conserved, yet Cryptic Epitope. *mBio* **2018**, *9*, e01674-18. [CrossRef] [PubMed]

86. Zhao, X.; Howell, K.A.; He, S.; Brannan, J.M.; Wec, A.Z.; Davidson, E.; Turner, H.L.; Chiang, C.-I.; Lei, L.; Fels, J.M.; et al. Immunization-Elicited Broadly Protective Antibody Reveals Ebolavirus Fusion Loop as a Site of Vulnerability. *Cell* **2017**, *169*, 891–904.e15. [CrossRef] [PubMed]

87. Janus, B.M.; van Dyk, N.; Zhao, X.; Howell, K.A.; Soto, C.; Aman, M.J.; Li, Y.; Fuerst, T.R.; Ofek, G. Structural Basis for Broad Neutralization of Ebolaviruses by an Antibody Targeting the Glycoprotein Fusion Loop. *Nat. Commun.* **2018**, *9*, 3934. [CrossRef] [PubMed]

88. Milligan, J.C.; Parekh, D.V.; Fuller, K.M.; Igarashi, M.; Takada, A.; Saphire, E.O. Structural Characterization of Pan-Ebolavirus Antibody 6D6 Targeting the Fusion Peptide of the Surface Glycoprotein. *J. Infect. Dis.* **2018**, *219*, 415–419. [CrossRef] [PubMed]

89. West, B.R.; Wec, A.Z.; Moyer, C.L.; Fusco, M.L.; Ilinykh, P.A.; Huang, K.; Wirchnianski, A.S.; James, R.M.; Herbert, A.S.; Hui, S.; et al. Structural Basis of Broad Ebolavirus Neutralization by a Human Survivor Antibody. *Nat. Struct. Mol. Biol.* **2019**, *26*, 204–212. [CrossRef] [PubMed]

90. Fan, P.; Chi, X.; Liu, G.; Zhang, G.; Chen, Z.; Liu, Y.; Fang, T.; Li, J.; Banadyga, L.; He, S.; et al. Potent Neutralizing Monoclonal Antibodies against Ebola Virus Isolated from Vaccinated Donors. *MAbs* **2020**, *12*, 1742457. [CrossRef] [PubMed]

91. Zhao, Y.; Ren, J.; Harlos, K.; Jones, D.M.; Zeltina, A.; Bowden, T.A.; Padilla-Parra, S.; Fry, E.E.; Stuart, D.I. Toremifene Interacts with and Destabilizes the Ebola Virus Glycoprotein. *Nature* **2016**, *535*, 169–172. [CrossRef] [PubMed]

92. King, L.B.; West, B.R.; Moyer, C.L.; Gilchuk, P.; Flyak, A.; Ilinykh, P.A.; Bombardi, R.; Hui, S.; Huang, K.; Bukreyev, A.; et al. Cross-Reactive Neutralizing Human Survivor Monoclonal Antibody BDBV223 Targets the Ebolavirus Stalk. *Nat. Commun.* **2019**, *10*, 1788. [CrossRef] [PubMed]

93. Beniac, D.R.; Booth, T.F. Structure of the Ebola Virus Glycoprotein Spike within the Virion Envelope at 11 Å Resolution. *Sci. Rep.* **2017**, *7*, 46374. [CrossRef] [PubMed]

94. Lee, J.E.; Kuehne, A.; Abelson, D.M.; Fusco, M.L.; Hart, M.K.; Saphire, E.O. Complex of a Protective Antibody with Its Ebola Virus GP Peptide Epitope: Unusual Features of a V Lambda x Light Chain. *J. Mol. Biol.* **2008**, *375*, 202–216. [CrossRef] [PubMed]

95. Olal, D.; Kuehne, A.; Bale, S.; Halfmann, P.; Hashiguchi, T.; Fusco, M.L.; Lee, J.E.; King, L.B.; Kawaoka, Y.; Dye, J.M.; et al. Structure of an Ebola Virus-Protective Antibody in Complex with Its Mucin-Domain Linear Epitope. *J. Virol.* **2011**, *86*, 2809–2906. [CrossRef]

96. Ilinykh, P.A.; Huang, K.; Santos, R.I.; Gilchuk, P.; Gunn, B.M.; Karim, M.M.; Liang, J.; Fouch, M.E.; Davidson, E.; Parekh, D.V.; et al. Non-Neutralizing Antibodies from a Marburg Infection Survivor Mediate Protection by Fc-Effector Functions and by Enhancing Efficacy of Other Antibodies. *Cell Host Microbe* **2020**, *27*, 976–991.e11. [CrossRef] [PubMed]

97. Mendoza, E.J.; Qiu, X.; Kobinger, G.P. Progression of Ebola Therapeutics During the 2014–2015 Outbreak. *Trends Mol. Med.* **2016**, *22*, 164–173. [CrossRef]

98. Lyon, G.M.; Mehta, A.K.; Varkey, J.B.; Brantly, K.; Plyler, L.; McElroy, A.K.; Kraft, C.S.; Towner, J.S.; Spiropoulou, C.; Ströher, U.; et al. Clinical Care of Two Patients with Ebola Virus Disease in the United States. *N. Engl. J. Med.* **2014**, *371*, 2402–2409. [CrossRef]

99. PREVAIL II Writing Group; Multi-National PREVAIL II Study Team; Davey, R.T., Jr.; Dodd, L.; Proschan, M.A.; Neaton, J.; Neuhaus Nordwall, J.; Koopmeiners, J.S.; Beigel, J.; Tierney, J.; et al. A Randomized, Controlled Trial of ZMapp for Ebola Virus Infection. *N. Engl. J. Med.* **2016**, *375*, 1448–1456.

100. Herbert, A.S.; Froude, J.W.; Ortiz, R.A.; Kuehne, A.I.; Dorosky, D.E.; Bakken, R.R.; Zak, S.E.; Josleyn, N.M.; Musiychuk, K.; Mark Jones, R.; et al. Development of an Antibody Cocktail for Treatment of Sudan Virus Infection. *Proc. Natl. Acad. Sci. USA* **2020**, *117*, 3768–3778. [CrossRef] [PubMed]

101. Dias, J.M.; Kuehne, A.I.; Abelson, D.M.; Bale, S.; Wong, A.C.; Halfmann, P.; Muhammad, M.A.; Fusco, M.L.; Zak, S.E.; Kang, E.; et al. A Shared Structural Solution for Neutralizing Ebolaviruses. *Nat. Struct. Mol. Biol.* **2011**, *18*, 1424–1427. [CrossRef] [PubMed]

102. Wec, A.Z.; Bornholdt, Z.A.; He, S.; Herbert, A.S.; Goodwin, E.; Wirchnianski, A.S.; Gunn, B.M.; Zhang, Z.; Zhu, W.; Liu, G.; et al. Development of a Human Antibody Cocktail That Deploys Multiple Functions to Confer Pan-Ebolavirus Protection. *Cell Host Microbe* **2019**, *25*, 39–48.e5. [CrossRef] [PubMed]

103. Xu, Y.; Roach, W.; Sun, T.; Jain, T.; Prinz, B.; Yu, T.-Y.; Torrey, J.; Thomas, J.; Bobrowicz, P.; Vásquez, M.; et al. Addressing Polyspecificity of Antibodies Selected from an in Vitro Yeast Presentation System: A FACS-Based, High-Throughput Selection and Analytical Tool. *Protein Eng. Des. Sel.* **2013**, *26*, 663–670. [CrossRef] [PubMed]

104. Oswald, W.B.; Geisbert, T.W.; Davis, K.J.; Geisbert, J.B.; Sullivan, N.J.; Jahrling, P.B.; Parren, P.W.H.I.; Burton, D.R. Neutralizing Antibody Fails to Impact the Course of Ebola Virus Infection in Monkeys. *PLoS Pathog.* **2007**, *3*, e9. [CrossRef]

105. Hastie, K.M.; Li, H.; Bedinger, D.; Schendel, S.L.; Dennison, S.M.; Li, K.; Rayaprolu, V.; Yu, X.; Mann, C.; Zandonatti, M.; et al. Defining Variant-Resistant Epitopes Targeted by SARS-CoV-2 Antibodies: A Global Consortium Study. *Science* **2021**, *374*, 472–478. [CrossRef]

MDPI

St. Alban-Anlage 66

4052 Basel

Switzerland

Tel. +41 61 683 77 34

Fax +41 61 302 89 18

www.mdpi.com

Pathogens Editorial Office

E-mail: pathogens@mdpi.com

www.mdpi.com/journal/pathogens